Handbook of
TEXTILE FIBRES

I. NATURAL FIBRES

Handbook of
TEXTILE FIBRES

By
J. GORDON COOK

BSc, PhD, CChem, FRSC

I. NATURAL FIBRES

WOODHEAD PUBLISHING LIMITED

Oxford   Cambridge   Philadelphia   New Delhi

Published by Woodhead Publishing Limited
80 High Street, Sawston, Cambridge CB22 3HJ, UK
www.woodheadpublishing.com; www.woodheadpublishingonline.com

Woodhead Publishing, 1518 Walnut Street, Suite 1100, Philadelphia, PA 19102-3406, USA

Woodhead Publishing India Private Limited, G-2, Vardaan House, 7/28 Ansari Road,
Daryaganj, New Delhi – 110002, India
www.woodheadpublishingindia.com

Formerly published by Merrow Publishing Co Limited
First published 1959
Second Edition 1960
Third Edition 1964
Fourth Edition 1968, reprinted 1974
Fifth Edition 1984, reprinted 1993
Reprinted by Woodhead Publishing Limited 2001, 2002, 2004, 2005, 2006, 2010, 2012

British Library Cataloguing in Publication Data
A catalogue record for this book is available from the British Library.

ISBN 978-1-85573-484-5 (print)
ISBN 978-1-84569-315-2 (online)

FOREWORD

The manufacture of textiles is one of the oldest and most important industries of all. Its raw materials are fibres, and the study of textiles therefore begins with an understanding of the fibres from which modern textiles are made.

In this book, an outline is given of the history, production and fundamental properties of important textile fibres in use today. The behaviour of each fibre as it affects the nature of its fabrics is discussed.

The book is in two volumes. Volume I deals with the natural fibres on which we depended for our textiles until comparatively recent times. Volume II is concerned with man-made fibres, including rayons and other natural polymer fibres, and the true synthetic fibres which have made such rapid progress in modern times.

The book has been written for all concerned with the textile trade who require a background of information on fibres to help them in their work. Every effort has been made to ensure that the text is accurate and up-to-date. The information on man-made fibres is based on facts supplied by the manufacturers of the fibres themselves.

In writing this book I have been given much encouragement and help by many individuals and organizations. The manufacturers of the man-made fibres mentioned in the text have gone to great trouble on my behalf in providing information and in checking the text before publication. I would like to acknowledge their help, with grateful thanks, and also that given to me by the following individuals and organizations:

D. A. Derrett-Smith, Esq., B.Sc., F.R.I.C., Linen Industry Research Association.

Dr C. H. Fisher, U.S. Dept. of Agriculture.

The Cotton Board, Manchester.

Stanley B. Hunt, Textile Economics Bureau.

Dr R. J. W. Reynolds, I.C.I. Dyestuffs Division.

Mr H. Sagar, I.C.I. Dyestuffs Division.

H. L. Parsons, Esq., B.Sc., F.R.I.C., Low and Bonar Ltd.

L. G. Noon, Esq., Wigglesworth and Co. Ltd.

Silk and Rayon Users Association.

J. C. Dickinson, Esq., International Wool Secretariat.

W. R. Beath, Esq., and his colleagues; Courtaulds Ltd.

E. Lord, Esq., B.Sc., Cotton, Silk and Man-Made Fibres Research Association.

K. J. Brookfield, Esq., Fibreglass Ltd.

F. H. Clayton, Esq., Wm. Frost Ltd.

Burlington Industries Inc.

J.G.C.

NOTE ON THE FIFTH EDITION

The text of this new edition of the *Handbook of Textile Fibres* has been revised and some sections have been rewritten to include newer information and broaden the coverage. Metric units as well as Imperial units are now used in order to bring the information into line with modern requirements.

I would like to thank again the many people who have kindly helped me in preparing the new edition, and I am especially grateful to the following firms and individuals who helped directly in the revision of the text of Volume 1.

Wigglesworth and Co. Ltd.; Mr V. J. Landon; Mr Albert E. Simons; Mr Jeremy Harris; Mr J. E. Malcolm.

Liverpool Cotton Services Ltd. (Cotton Outlook); Mr John Garner.

U.S. Department of Agriculture (Memphis); Mr Donald W. Bratton.

International Wool Secretariat; Mr P. C. Marshall and colleagues.

The Wool Bureau Inc.; Mr R. C. Freeman.

H. T. Gaddum Ltd.; Mr P. W. Gaddum.

J.G.C.

CONTENTS

INTRODUCTION

FIBRES FOR CLOTHES

During the early days of his existence, man depended upon animal skins and furs to keep him warm. But as the years passed, his susceptibilities became more tender and his hide less coarse. A sheepskin wrapped carelessly round the body may be better than nothing for keeping out the cold – but only just. Inflexible and uncomfortable, it would not fit *homo sapiens* as well as it had the sheep.

Inevitably, man began to look around for something that would keep him warm more elegantly and more comfortably than an evilsmelling hide. At some point in history, he found that the long thin fibres produced by plants and animals could be twisted together to form a thread. These threads could then be interlaced to provide a flexible, warm and supremely comfortable material such as he had never known before. He had discovered cloth.

Natural Fibres
Since those early days, all the hundreds of different sorts of natural fibre have been collected and examined as potential raw materials for cloth. The hairs of animals like the sheep; the 'stringy' portions of plants ranging from the coarse backbone of the nettle stem to the fine seed-fibre of the cotton plant; the delicate filaments formed by insects and other creatures, like the spider's web or the cocoon of the silkworm – all these can be twisted together to form a thread and then woven into cloth.

In their basic properties – their fineness and flexibility, their resilience and shape – these natural fibres vary widely, and the types of cloth they provide are correspondingly diverse. But in general, cloth possesses certain common and altogether satisfying characteristics that have made it into one of the essential materials of our modern world. Cloth is strong and yet sufficiently supple to take up the peculiar contours of the human body. Made into clothing it is hard-wearing and yet permits free movement; it is warm, but at the same time does not seal up the body and prevent its breathing.

As textiles have developed over many hundreds of years, the most suitable natural fibres have been selected and have become the basis of the textile industries of the world. Today, as for many years past, cotton, wool, jute, flax and silk are the most important of our natural textile fibres, with a number of others doing relatively unimportant jobs.

Upon these few natural fibres, man has depended for centuries for the clothes that keep him warm. Prior to the Industrial Revolution, spinning and weaving were accepted routines of daily life in every home, and the making of fibres into textiles remained a household industry. It was a craft, a skilled occupation of working people who handed down its secrets from one generation to the next.

Spinning

The first fundamental process in making cloth is spinning, which is one of the oldest industrial arts in the world. Spinning converts a mass of short fibres into long threads or yarns suitable for weaving together into cloth. For thousands of years, the techniques of spinning natural fibres altered little. Hand-spinning was a process that depended on manual dexterity and skill, using the very simplest of mechanical devices. It is only in the last two hundred years that spinning machines have transformed the production of yarns and threads into a mechanized industry.

Though hand-spinning developed simultaneously in many parts of the world, and many different fibres were used, the simple spinning instrument – the spindle – is basically the same no matter where it comes from. The spindle is little more than a piece of smooth, rounded wood about a foot long, tapered at each end and weighted with a little disc of wood, clay or stone in the middle. The latter acts as a flywheel, giving momentum and stability to the spindle as it rotates.

At one end of the spindle there is a little notch in which the fibres are caught up as they are being twisted into yarn. The spindle is rotated by being rolled against the leg, or it is simply twirled between the fingers. As the twisted yarn is formed, more fibres are pulled out steadily from the mass of wool or cotton or other fibre until a length of yarn has been produced. This stretch of yarn is then wound up onto the spindle shaft and the process is repeated.

In this way, the simple hand spindle was used for centuries to produce continuous lengths of yarn from the masses of animal or vegetable fibres that nature made available. In the hands of an expert craftsman, the spindle could – and indeed still can – provide yarns of incredible delicacy and uniformity. The wonderful Indian muslins were once made entirely from cotton spun so fine on hand spindles that 28g (1 oz) of fibre would make 15m (50 ft) of yarn. The spindle was a thin length of bamboo cane weighted with a small disc of clay.

By the fourteenth century, the first steps towards mechanization had been taken. In India and Europe, the spinning wheel had been developed. This was a simple instrument in which the spindle was mounted horizontally in a wooden frame. The little flywheel in the centre of the spindle was converted into a pulley by cutting a groove in the rim. A leather thong was passed round this pulley and round a larger wheel; when the larger wheel was turned with the left hand, the spindle rotated much faster than it could be spun directly in the hand. Fibres were fed to it by the right hand of the spinner, and twisted into yarn in the usual way. When an arm's length had been made, the yarn was held at right angles to the spindle and wound up onto it. Then a further arm's length was made, and so on until a bobbin of yarn had been formed on the spindle shank.

By the sixteenth century the spinning wheel had been fitted with a treadle, leaving the operator with both hands free for manipulating the threads.

Weaving

The yarns made with these simple manually-operated spinning machines were converted into cloth by being woven on hand-looms. The loom is a device in which yarns can be criss-crossed or interlaced and so built up into a textile fabric. The simplest of all fabrics is a plain cloth in which rows of threads at right-angles to one another pass over and under each other alternately. Cloth of this sort is made on a hand-loom in which every alternate thread running lengthwise (the warp) is lifted and the remaining threads are lowered. A transverse thread (the weft or filling) is passed between the two sets of warp threads, whose positions are then reversed. In this way, the transverse thread is locked in place, passing over and under alternate threads in the warp. All manner of variations can be superimposed upon this simple theme, enabling the skilled weaver to create an immense variety of intricate weaves in his fabric.

Hand-loom weaving was the second stage in the production of cloth by village communities. The weaver was a local craftsman who enjoyed a status similar to that of the carpenter or the blacksmith.

Textile manufacture thus became a trade based on traditional experience and skill. It remained an occupation of the ordinary people, who had neither the time nor opportunity to pay attention to the fundamentals of their trade. Pure science before the days of steam was seldom linked with mundane occupations such as the processing of fibres into textiles.

During the eighteenth century, the Industrial Revolution swept over Britain as steam began to turn the wheels of industry. And what trades could be better served by steam than spinning and weaving? Steam engines took over from the water-wheels in mills that had been built beside the Pennine valley streams. Gradually, the spinning wheels and looms were abandoned in the cottages as their owners migrated to the great steam-powered mills in the valleys of Lancashire and in the Yorkshire dales. Throughout the nineteenth century, Britain's textile industry thrived and became the mainstay of the country's trade. Steam power allied with engineering invention transformed the old crafts into an industry whose market was the world.

Scientific Renaissance

This same century was to see a renaissance of science in Britain which followed John Dalton's development of the atomic theory in 1808. But science was slow to reach the textile trade; the manufacturers were too busy turning out their goods to worry unduly over the whys and wherefores of their raw materials. Nature had provided cotton and wool, flax and silk; surely it was now up to man to make the best use of them that he could?

So, during most of the nineteenth century, textile progress lay not so much in the integration of the industry with advancing scientific knowledge, but in the continued application of inventive and engineering skill to the spinning and weaving processes.

It was on this basis that Britain's textile industry built itself to greatness in Victoria's reign; and even by the beginning of the present century knowledge of the fundamental structure of the textile fibres remained scanty and uncertain.

It is only during the last fifty or sixty years that science has really begun to play a major role in the textile industry. As we have learned something of the chemistry and physics of textile fibres, so we have been able to create a range of completely new fibres which have changed the entire outlook of the textile trade. Rayon, nylon and other man-made fibres are being manufactured in enormous quantities and nature's monopoly of textile fibre production has been broken.

Today, the importance of research and scientific understanding in the textile industry is established. The fibres on which the entire industry is based are the subject of a vast amount of academic and industrial research. Textile progress is no longer dependent simply on inventiveness and engineering skill; textile manufacture has become a modern scientific industry that must keep abreast of scientific progress and discovery.

The 'big four' – cotton, wool, flax and silk – are still used more extensively than any other natural textile fibres. But the manufacture of rayon and synthetic fibres has attained the status of a major world industry, and output is increasing. Moreover, the discovery of nylon stimulated research on synthetic fibres which has given us a range of synthetic fibres which increases year by year.

What is a Textile Fibre?

The use of textiles for clothing and furnishing depends upon a unique combination of properties. Textiles are warm; they are soft to the touch; they are completely flexible and thus take up any desired shape without resistance; and they are usually hard-wearing.

The reason for these properties is to be found in the structure of textile materials. Textiles are derived from threads or yarns which have been interlaced in one way or another. The threads themselves are flexible, and in their loose interweaving they remain flexible, conferring this property on the cloth itself.

In their turn, the threads or yarns are built up by twisting together the long, thin, flexible but strong things we call fibres. Ultimately, therefore, the properties of any material must depend very largely on the properties of the fibres from which it is made. The spinning and weaving processes obviously have their effect on the final textile. A worsted suit, for example, bears little superficial resemblance to a baby's cardigan, though both are made from wool. But the basic natures of the two garments are similar, and are a consequence of the fact that each is made from wool.

For a fibre to be suitable for textile purposes, certain qualities are desirable; others are essential. First, to be a fibre at all, the length must be several hundred times the width. It is this that enables fibres to be twisted together to form a yarn or thread.

In addition, the fibre must be strong and yet extremely flexible. Strength is needed to enable it to withstand the spinning and weaving processes, and to provide strength in the final cloth. Flexibility permits the fibres to be spun and woven, and gives to a textile its unique draping characteristics.

The actual length of the fibre is important. It can be infinitely long, but should not be shorter than 6 – 12mm (¼ – ½ in), or it may not hold together after spinning. The width of the fibre can vary between considerable limits, and it is upon this that the fineness of the material eventually depends. Silk, for example, is a fine fibre and yields a delicate cloth; jute is a coarse fibre that is largely used for making sacks.

In addition to having strength and flexibility, a textile fibre should be elastic. Brittleness leads to poor wear in the garment; elasticity allows the material to 'give' when subjected to a stretching force.

Waviness, or crimp, is a natural feature of certain fibres such as wool. It affects the 'holding together' power of the fibres in the spun yarn and controls the porosity and warmth of the fabric.

The ability of a fibre to absorb moisture influences the hygienic qualities of the cloth. Fibres that cannot absorb moisture may help to make the cloth feel clammy when it is worn.

The weight of a fibre affects the draping qualities when it is made into a cloth. If the fabric is too light, it may not drape well; yet if it is too weighty, the material will be heavy and dull.

With all the variability possible in these important properties, it is not surprising that we find such diverse characteristics in the natural fibres. Nor is it reasonable to expect that anything that looks fibrous will be suitable for making into textiles.

When we add to these requirements the essentials of abundance and cheapness, we find that the number of fibres suitable for large-scale textile use has narrowed down to relatively few.

Some of them, like cotton and flax, are vegetable fibres which nature uses for some essential purpose in the growing plant; others, like wool or silk, are produced by the animal world.

Classification*

The fibres used in modern textile manufacture can be classified into two main groups (a) natural and (b) man-made fibres. The natural fibres are those, such as cotton, wool, silk and flax, which are provided by nature in a ready-made fibrous form. The man-made fibres, on the other hand, are those in which man has generated a fibre for himself from something which was not previously in a suitable fibrous form.

* See chart on page xxvii.

NATURAL FIBRES can be subdivided into three main classes, according to the nature of their source.

    (a) Vegetable fibres
    (b) Animal fibres
    (c) Mineral fibres

Vegetable fibres include the most important of all textile fibres – cotton – together with flax, hemp, jute and other fibres which have been produced by plants. They are based on cellulose, the material used by nature as a structural material in the plant world.

Animal fibres include wool and other hair-like fibres, and fibres, such as silk, produced as filaments by cocoon-spinning creatures. These animal fibres are based on proteins, the complex substances from which much of the animal body is made.

Mineral fibres are of limited importance in the textile trade. Asbestos is the most useful fibre of this class; it is made into special fire-proof and industrial fabrics.

MAN-MADE FIBRES can be sub-divided into two distinct classes, according to the source of the fibre-forming substance from which they are made.

    (a) Natural polymer fibres
    (b) Synthetic fibres

Natural polymer fibres are those in which the fibre-forming substance has been made by nature. Vast quantities of cellulose, for example, are available to us in the plant world. Only a small fraction of this cellulose is used by nature for making fine fibres such as cotton. Most of it is used as a structural material, for example in the trunks of trees and the skeleton framework of stems and leaves. This cellulose is largely useless to us as a direct source of textile fibres; it is in fibrous form, but is contaminated with other substances.

In the last half-century or so, we have learned how to manipulate this natural cellulose into a form suitable for use as textile fibres. It is the source of the fibres which became known as artificial silks.

In these natural polymer fibres nature has done the work of creating a substance (cellulose) capable of taking on a fibrous form. Man has merely taken a further step by using this cellulose as raw material for a fibre.

In a similar way, it has been possible to use materials made by animals as a source of man-made fibres. The proteins used for so many structural purposes in the animal world are often capable of dyeing techniques in detail. Dyeing is a complex and highly skilled art, and detailed procedures are described in other books.

(c) STRUCTURE AND PROPERTIES

This section summarizes the important characteristics of a fibre under a series of sub-headings.

(1) FINE STRUCTURE AND APPEARANCE. The surface structure of a fibre is most important in that it controls the behaviour of the fibre in the yarn or fabric. The rough scaly surface of wool, for example, influences the felting and shrinkage properties of wool fabrics, and helps to give wool its characteristic handle. The scales enable individual fibres to grip one another when twisted together as a yarn. The convolutions of the cotton fibre, similarly, enable fibres to grip one another when spun.

The smooth, glassy surface of a fibre such as nylon, on the other hand, affects the lustrous appearance of the fabric. Smooth surfaces may not cling to dust and dirt so readily as rough surfaces do.

The cross-sectional shape of a fibre has an important influence on its behaviour in a textile fabric. Fibres of circular or near-circular cross-section often have an attractive handle. Wool, for example, is a fibre of near-circular cross-section; it has a more 'comfortable' feel than cotton which has a flatter, ribbon-like cross-section. 'Orlon', on the other hand, which has a dog-bone cross-section, has a very good handle.

Circular fibres often have a poorer covering-power than the flatter fibres.

Diagrams showing the microscopic appearance (cross-sectional and longitudinal) are provided for many fibres.

(2) TENSILE STRENGTH. This is the breaking strength of any material, which is commonly expressed as force per unit cross-sectional area, e.g. as dynes per square cm. In these terms, we may describe the ability of a bundle of fibres, or a yarn, to resist breakage under tension.

When a single fibre is being considered, the strength of the fibre is commonly described as *tenacity*, which is a measure of *specific stress* at break,

$$\text{i.e. } \frac{\text{breaking load}}{\text{mass per unit length}}$$

Tenacity is expressed in terms of grams per decitex or centinewtons per tex (cN/tex).

Two fibres with identical tenacities may have different tensile strengths; if their densities are different, the cross-sectional areas will be different too.

(3) ELONGATION. When a fibre is subjected to a force, it will stretch to a certain degree. This stretching is described as elongation or extension, in terms of a percentage of the fibre's original length. It can be measured either as an elongation under a certain load, or as the elongation reached when the fibre breaks. Unless specified to the contrary, the figure given represents the elongation at break.

(4) ELASTIC PROPERTIES. *Elastic Recovery*. When a fibre is stretched by a small amount, it may exhibit almost perfect elasticity. That is to say, it will return to its original length when it is released. The *elastic recovery* in this case is 100 per cent. If, however, the fibre is subjected to a greater degree of stretch, it may react in a much more complex way. Some permanent deformation may take place, so that when it is released the fibre will return to an elongated form. It recovers from some of its elongation, but not all.

This behaviour of a fibre can be denoted by describing its elastic recovery at certain elongations (specified as percentage of original length). Thus, in the case of a fibre which returns completely to its original length after, say, a 2 per cent elongation, we can say that the elastic recovery is 100 per cent at 2 per cent elongation. In the case of a fibre which retains half its extra length after release from an 8 per cent elongation, we say that it has a 50 per cent elastic recovery at 8 per cent elongation.

The elastic properties of a fibre are normally defined only with limited usefulness in this way. The recovery of a fibre, for example, depends upon the length of time it is held in the stretched position. Also, the degree to which it recovers depends on the time between its release from tension and the taking of the measurement.

Stress-Strain Diagram

The tensile and elastic properties of a fibre are usually summarized in a *stress-strain diagram*. In this diagram, the strain (i.e. the distortion in the fibre) is plotted against the stress (i.e. force) exerted on the fibre. A stress-strain diagram gives a much more complete record of the behaviour of a fibre under tension than isolated figures can.

Typical stress-strain diagrams are provided for many fibres.

A straight line on the stress-strain diagram may indicate that the fibre is truly elastic. The extension of the fibre is proportional to the applied load. This is, however, rarely achieved in practice. As the load on a fibre increases beyond that needed to cause a few per cent extension, the deformation of the fibre is greater than that due to true elasticity. Superimposed upon the 'elastic' stretch there is some more or less permanent deformation of the fibre, or plastic flow.

As the tension increases, the stress-strain curve indicates how the fibre continues to deform up to the point at which it eventually breaks.

The stress-strain diagram therefore provides a much more complete picture of the deformation caused in a fibre as tension is applied to it. The diagram includes tenacity and elongation at break. Elastic recovery and the slow recoverable deformation described as 'creep' are determined from a number of stress-strain diagrams where repeated stresses are given, and the return paths measured.

The stress-strain behaviour of a fibre is of great importance in practice, and influences to a large degree the behaviour of the fibre in textile manufacture. During processing of the fibre into a yarn and weaving of the yarn to fabric the fibres are under varying degrees of tension. They should be able to withstand these tensions without stretching permanently to any great degree.

Wool is unusual in that it can stretch by 35 per cent and will return to its original length when relaxed. Cotton, on the other hand, has an extension at break of only about 5–10 per cent.

The general reaction of a fibre to longitudinal tensions and to flexing backwards and forwards has an immense influence on the properties of the cloth made from the fibre. A resilient fibre such as wool will tend to return to its original shape after a fabric has been crushed or creased. The crease-resistance of a fabric is usually a consequence of the resilience of the fibre itself.

Work of Rupture. The area below the stress-strain curve provides a measure of the energy needed to break the fibre. It indicates the ability of the fibre to withstand sudden shocks, and is measured in grams per decitex or centinewtons per tex.

Initial Modulus. This is a measure of a fibre's resistance to small extensions. A high modulus means that the fibre has a good resistance to stretching, and a low modulus means that it requires little force to stretch it. Flexibility and modulus are closely linked, a low-modulus fibre tending to be flexible, and a high-modulus fibre tending to be brittle.

Average Stiffness. This is the ability of a fibre to carry a load without deformation. It is based on the modulus of elasticity, and is expressed as grams per dtex or cN per tex.

Average Toughness. This is the ability of a fibre to endure large permanent deformations without rupture. It is expressed as grams per dtex or cN per tex.

(5) SPECIFIC GRAVITY. This is a measure of the density of a fibre; it is the ratio of the mass of a material to the mass of an equal volume of water at 4°C. This is an important characteristic of any fibre; it affects the way in which a fabric will drape.

(6) EFFECT OF MOISTURE. All fibres tend to absorb moisture when in contact with the atmosphere. The amount absorbed depends upon the relative humidity of the air.

In practice, the moisture-absorbing properties of a fibre are described by a figure known as the 'regain'. This is the weight of moisture present in a textile material expressed as a percentage of its oven-dry weight (i.e. the constant weight obtained by drying at a temperature of 105 to 110°C.).

The 'percentage moisture content' of a fibre is the weight of moisture it contains, expressed as a percentage of the total weight. This is a measure of the amount of water held under any particular set of circumstances.

Fibres vary greatly in the amount of moisture they will absorb. Wool, for example, has a regain of 16 per cent, acetate of 6 per cent and 'Dynel' 0·4 per cent. A fibre which absorbs water readily is often most suitable for use in certain types of clothing fabrics. These fabrics will absorb perspiration from the body and will hold considerable amounts of water without feeling clammy. The ability of a fibre to absorb moisture will also affect the processing and finishing of yarns and fabrics. Dyestuffs are generally able to penetrate a moisture-absorbing fibre much more easily than they will penetrate a fibre that does not absorb much moisture.

The new synthetic fibres, which often have a very low moisture regain, are easily washed and dried by comparison with fibres which absorb a lot of moisture. On the other hand, they tend to accumulate charges of static electricity much more readily than the moisture-absorbing fibres.

The tensile properties of a fibre are affected significantly by the water it absorbs. A fibre which absorbs water freely will usually suffer a loss in tensile strength when wet. (Cotton is an exception.) Elongation at break is also increased.

As fibres absorb moisture they may swell to a considerable degree.

(7) THERMAL PROPERTIES. All fibres are affected in one way or another as they are heated. Some, like wool, will begin to decompose without melting; others, like polyethylene or acetate will soften and melt before decomposition sets in. The behaviour of fibres on heating is of real importance, particularly within the range of temperatures that are met in practical use. Fabrics should, for example, withstand the temperatures used in laundering and ironing without undue deterioration.

Many of the new synthetic fibres are thermoplastic substances; that is to say, they will soften as they are heated. The temperature at which they soften largely determines their practical usefulness in the textile field.

In the presence of air, most fibres will burn. The readiness with which they catch fire and support combustion is of immense importance. Many accidents are caused every year by clothing catching fire, and there is an increasing realization of the need for reducing the flammability of textile fibres and fabrics.

(8) EFFECT OF SUNLIGHT. Almost every fibre is affected by the powerful radiations of sunlight. Some will decompose and deteriorate fairly rapidly, losing tensile strength and changing colour. Others will resist deterioration for years, and are particularly useful for fabrics such as curtains, awnings and furnishings which are constantly exposed to light.

(9) CHEMICAL PROPERTIES. Modern techniques of processing fibres, yarns and fabrics often involve the use of chemicals in great variety. Bleaching agents, detergents, alkaline scouring agents, dyeing assistants and other chemicals are used in preparing the finished textile. The fibre itself must be able to withstand these substances without suffering harmful effects.

(10) EFFECT OF ACIDS. Textiles are commonly subjected to acid

solutions of one sort or another, and the effects of different acids under varying conditions are important.

(11) EFFECT OF ALKALIS. From the very earliest times, alkaline agents have been used for washing and scouring textiles. Soap itself forms an alkaline solution in water.

(12) EFFECT OF ORGANIC SOLVENTS. The introduction of dry-cleaning has made solvent-resistance of great importance in a textile. Solvents such as carbon tetrachloride and trichloroethylene are commonly used for cleaning fabrics, and the effect of these solvents on the fibre itself is obviously important.

(13) RESISTANCE TO INSECTS. The cellulose of plant fibres and the protein of wool and other animal fibres are substances produced by living things. They are, as might be anticipated, enjoyed by other living things as food.

Wool suffers more than other fibres from the fact that it is eaten by certain types of moth grub and beetle. Many fibres, particularly the synthetics, are not attacked in this way.

(14) RESISTANCE TO MICRO-ORGANISMS. Cellulose is attacked by certain moulds and bacteria, which decompose it and make use of the degradation products as food. Textiles stored in damp warehouses are often affected by mildews, which may discolour and weaken the fibres to the point at which they become useless.

(15) ELECTRICAL PROPERTIES. The dielectric strength of a fabric is important if the material is to be used for insulation purposes in the electrical industry. It also influences the degree to which static electricity will accumulate on a yarn or fabric during processing or wear. Static electricity may be produced by friction between the yarns or fabrics and the surfaces they meet on processing machinery. The electricity often causes serious difficulties by entangling or misaligning yarns on machinery and attracting dust and fluff to the finished fabric.

The electrical resistance of a fibre may be described in terms of the mass specific resistance, i.e. the resistance of a 1 gram specimen 1 cm. long.

The production of static electricity is affected greatly by the moisture-absorbing characteristics of the fibre. A damp fibre will conduct electricity away as it is formed, so that pools of static do not collect on the fibre.

These properties can be regarded as fundamental characteristics of a fibre, and they are discussed in the 'Structure and Properties' section of each important fibre.

(d) THE FIBRE IN USE

In this section, the influence of the properties of the fibre on its behaviour in practical use is considered.

FIBRE CLASSIFICATION CHART

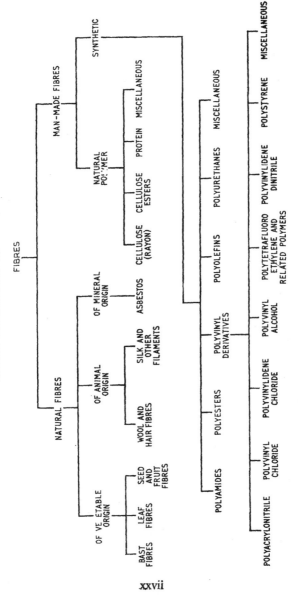

NATURAL FIBRES

A: OF VEGETABLE ORIGIN

B: OF ANIMAL ORIGIN

C: OF MINERAL ORIGIN

A: NATURAL FIBRES OF VEGETABLE ORIGIN

Introduction

In the complex designs and structures of the higher plants, nature has used fibres as the basis of the strength-providing skeleton. Bundles of fibres, bound together by natural gums and resins, run through the roots and stems and leaves of plants. Some of these fibrous structures act as pillars and girders, for example in the woody cores of the trunks and branches of a tree. Others function as hawsers, like the fibrous bundles that take the strain in the stalks and stems of less robust plants, or in the roots that grip the ground and hold the plant firm against the buffeting of the weather.

Flimsy fibres, delicate and yet supremely flexible and strong, are used by many plants as streamers to catch the wind and carry their seeds for immense distances through the air.

These vegetable fibres are all based upon cellulose,* the substance related to the starch and sugars which the plant builds up from water and from carbon dioxide gas absorbed through its leaves. The resources of cellulose fibre available to us in the plant world are virtually inexhaustible. But only a comparatively small proportion of these resources can be made use of directly as textile fibres. The strands of cellulose fibre in plants are associated with varying amounts of other natural substances such as lignin, pectins, hemicelluloses, waxes and gums. The amount of these associated substances and the ease with which the cellulose fibre can be separated from them determine how useful any vegetable fibre can be as a textile material.

The cellulosic fibres at present in use as textile raw materials can be classified most conveniently by referring to the part of the plant from which they come. There are three main groups:

(1) THE BAST OR STEM FIBRES, which form the fibrous bundles in the inner bark (phloem or bast) of the stems of dicotyledenous plants (i.e. plants which form two seed-leaves).

(2) THE LEAF FIBRES, which run lengthwise through the leaves of monocotyledenous plants (i.e. plants which form one seed leaf).

(3) THE FIBRES OF SEEDS AND FRUITS, including the true seed-hairs and the flosses.

* See page 75.

3

THE BAST FIBRES

The bast fibres form bundles or strands that act as hawsers in the fibrous layer lying beneath the bark of dicotyledenous plants. They help to hold the plant erect.

These fibres are constructed of long thick-walled cells which overlap one another; they are cemented together by non-cellulosic materials to form continuous strands that may run the entire length of the plant stem.

The strands of bast fibres are normally released from the cellular and woody tissue of the stem by a process of natural decomposition called retting (controlled rotting). Often, the strands are used commercially without separating the individual fibres one from another.

On a tonnage basis jute is the most important of all the bast fibres; the world output (about 4 million tonnes in 1979) is greater than that of all the other bast fibres combined. But most of the world's jute is made into sacking and baggage cloths.

The production of flax is roughly one seventh that of jute (606,000 tonnes in 1979). But flax is the fibre from which we make linen; it is on this basis the most important of the bast textile fibres.

FLAX

Flax was probably the first plant fibre to be used by man for making textiles, at least in the Western hemisphere. Specimens of flax have been found in the prehistoric lake dwellings of Switzerland, and in the tombs of Ancient Egypt. The evidence of biblical writings shows that the spinning and weaving of flax were well advanced thousands of years ago. Linen mummy-cloths have been identified as more than 4,500 years old.

Flax fibre comes from the stem of an annual plant *Linum usitatissimum*, which grows in many temperate and sub-tropical regions of the world. In the inner bark of this plant there are long, slender, thick-walled cells of which the fibre strands are composed.

From the Mediterranean region, flax-growing spread over Europe. Centuries before the beginning of the Christian era, Phoenician traders were bringing Egyptian linen to Britain. Roman legions carried the Mediterranean textile skills, including the crafts of spinning and weaving flax, to every corner of their empire.

During the seventeenth century, linen manufacture became established as a domestic industry in many countries of Western Europe. Flax from Germany was the raw material for a flourishing linen industry that grew in the Low Countries. Linen manufacture spread from Western Europe into England, Scotland and Ireland, stimulated by the flow of French and Flemish weavers who were driven from their homes by religious persecution.

Until the seventeenth century only small amounts of flax were grown in England. Competition from wool had stifled the linen industry. The arrival of linen workers from France and the Low Countries created a demand for flax that was met by importation of the fibre, largely from Russia.

Irish Linen Industry

In Ireland, official encouragement by the English Government led to the growth of a flourishing linen industry. But during the eighteenth century, the inventions of Arkwright, Hargreaves and Crompton were developed to the almost exclusive benefit of the cotton industry. And with the rise of cotton, the linen industry was forced into the background. The domestic spinning of flax, and the hand-loom weaving of linen began to diminish. Gradually linen manufacture retired into one or two strongholds, such as Northern Ireland, where it has survived to the present day.

In 1810, Napoleon I offered a reward of a million francs to the man who could devise a machine for spinning flax. A few weeks later, Philippe de Girard patented a machine to do the job, but he did not collect his reward. Five years later, he was invited by the Austrian Government to establish a spinning mill near Vienna; this ran for a number of years but was not a commercial success.

Meanwhile, in Britain, John Kendrew and Thomas Porthouse, of Darlington, patented a flax-spinning machine that was to form a basis for our present spinning machines. But, compared with cotton, progress was slow.

Today, the flax plant is grown for its fibre mainly in Europe, including the U.S.S.R.; elsewhere it has become of only limited importance as a source of fibre. In 1973 Russia grew about three-quarters of the world output of around 600,000 tonnes.

Since the war, accurate figures have not been available from some countries, but it is estimated that flax and tow production in 1977 was some 580,000 tonnes. Of this, France grew 75,000 tonnes, Belgium 14,000 tonnes, the Netherlands about 9,000 tonnes.

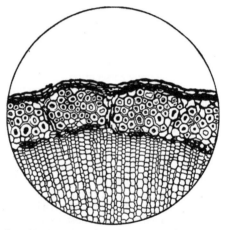

Flax. This section of the stem of a flax plant shows the bundles of fibre cells lying below the surface layer.

As a source of linseed oil, the flax plant is grown extensively in Canada, U.S.A. and Argentina, but the output of fibre from these countries is relatively insignificant. Soviet Russia grows flax for both fibre and oil.

PRODUCTION AND PROCESSING

Grown for fibre, the flax plant is an annual that reaches 90 – 120cm

(3 – 4 ft). It has a single slender stem that is devoid of side branches other than those which bear the flowers. When the plants have flowered and the seeds are beginning to ripen, the crop is pulled up by the roots (by hand or by mechanical pullers). About one-quarter of the stem consists of fibre.

Retting

The flax fibres are held together in the stems by woody matter and cellular tissue, and 'retting' is a fermentation process that frees the fibres from these materials.

Retting may be carried out in one of several ways:

(1) DAM-RETTING. The flax plants after pulling are tied up in sheaves or 'beets' and immersed for about ten days in water in special dams or ponds dug in the ground. An obsolete method no longer practised except in Egypt.

(2) DEW-RETTING. The crop is spread on the ground after pulling and left for several weeks. Wetting by dew and rain encourages fermentation by moulds to take place.

Dew-retting tends to yield a dark-coloured fibre. It may be used in regions where water is in short supply; it is commonly practised in the U.S.S.R. and France. This is the method by which some 85% of the West European crop is retted. It is far less labour intensive than water-retting and therefore less expensive.

(3) TANK-RETTING. After harvesting, the seed bolls are stripped from the stems by reciprocating metal combs. The de-seeded straw, tied in bundles, is packed into concrete tanks which are filled with water and artificially heated to about 30°C. Retting is completed in about three days. Some of the best and most uniform fibre is produced by this process. Almost all of the flax from the Courtrai district of Belgium is tank-retted. The highest quality of straw may be double-retted, i.e. the partly retted straw is removed from the tank, dried, and then given a further period of retting.

Flax is brought in from a large area to be retted centrally in this way. The advantage of this form of retting lies in the fact that conditions can be controlled and the process can be carried out at any time of the year. In the Belgian method the straw is usually given a preliminary steeping treatment.

Flax used to be retted in the River Lys in Courtrai by immersing the straw in wooden crates, but retting in the Lys is no longer permitted in Belgium.

(4) CHEMICAL RETTING. Retting can also be carried out by treating the flax straw with chemical solutions. Such reagents as caustic soda, sodium carbonate, soaps and dilute mineral acids have been employed with some success. In general, chemical retting of the straw proved to be a more costly process than biological retting and the fibre produced was no better. More recently, attention has been turned towards the chemical treatment of fibre extracted in the green state from unretted straw. With developments in chemical plant for the processing of fibre, this method becomes an economic possibility. A third alternative is to prepare the unretted fibre into rove, and boil or bleach the rove before spinning. During the last war, many thousands of tonnes of green fibre were spun from boiled rove in this way.

'Cottonization' of flax is a form of chemical retting which is carried to the point where the flax is separated into very fine strands. The flax can then be spun on cotton-spinning machinery. This was carried out in Germany and other Continental countries during the war.

Breaking and Scutching

The next stage in fibre-production is 'breaking'. The straw is passed between fluted rollers in a breaking machine, so that the woody core is broken into fragments without damaging the fibres running through the stems. The broken straw is then subjected to the process known as 'scutching', which separates the unwanted woody matter from the fibre. This is done by beating the straw with blunt wooden or metal blades on a scutching machine. The woody matter is removed as *shive*, which is usually burnt as fuel, leaving the flax in the form of long strands formed of bundles of individual fibres adhering to one another.

Hackling

After scutching the fibres are usually combed or 'hackled' by drawing them through sets of pins, each successive set being finer than the previous one. The coarse bundles of fibre are, in this way, separated into finer bundles, and the fibres are also arranged parallel to one another. The long fine fibres are known as *line*: the shorter fibres or *tow* are spun into yarns of lower quality.

The tow is subjected to further combing or 'carding', which aligns the fibres more accurately alongside one another. They are then collected into the loosely-held rope of fibre called a *sliver* or *rove*.

Spinning

Spinning may be carried out dry for the coarse yarns and wet for the finer yarns. In wet spinning the rove is passed through a bath of hot water. This softens the gummy matter holding the strands together, enabling them to be drawn out (drafted) and aligned more perfectly as the rove is elongated and twisted during spinning.

Dyeing

Most linens are dyed in the piece, using techniques similar to those for cotton. Good quality dyestuffs are generally used, as flax is an expensive fibre and linen is a high-quality textile. Vat dyes are used extensively for linen, giving excellent fastness to light and washing.

Azoics and sulphur colours are also used, and selected direct colours with good light fastness are used for furnishing fabrics.

The flax fibre is harder than cotton, and dyestuffs do not penetrate so readily into it. Special techniques are used during linen dyeing to ensure maximum penetration of the dye, such as the pigment padding process (vat dyeing).

STRUCTURE AND PROPERTIES

Flax fibre strands in the scutched state vary in length from a few centimetres (tow fibre) to as much as 1 metre (line). A good fibre averages 45 – 60cm (8 – 24 in). By the time the fibre reaches the spinning stage is has been broken down in length. Even the fibres in line yarn may be shorter than 30 – 38cm (12 – 15 in).

Commercial flax is in the form of bundles of individual fibre cells held together by a natural binding material. Scutching and hackling tend to break up the coarse bundles of fibre as they exist in the bast, but do not separate the fibre strands into their individual fibre cells.

Flax is usually coloured yellowish-white, but the shade of the raw fibre varies considerably depending upon the conditions under which it has been retted. Dew-retted fibre is generally grey.

Flax is usually soft and has a lustrous appearance. The lustre improves as the flax is cleaned, wax and other materials being removed.

The highest quality flaxes come from Belgium, Northern France and the Netherlands. Russian flaxes are generally weaker but are remarkable for their fineness of fibre.

Fine Structure and Appearance

Strands of commercial flax may consist of many individual fibre cells; they vary in length from 6 – 65mm (¼ – 2½ in) with a mean diameter of about 0.02mm (1/1200th in). Seen under the microscope, the fibre cells show up as long transparent, cylindrical tubes which may be smooth or striated lengthwise. They do not have the convolutions which are characteristic of cotton. The width of the fibre may vary several times along its length. There are swellings or 'nodes' at many points, and the fibres show characteristic cross-markings.

The fibre cell has a lumen or canal running through the centre; the lumen is narrow but clearly defined and regular in width. It disappears towards the end of the fibre, which tapers to a point.

The cell walls of the flax fibre are thick and polygonal in cross-section.

Immature flax fibres are more oval in cross-section, and the cell walls are thinner. The lumen is relatively much larger than in the mature fibre.

Tensile Strength

Flax is a stronger fibre than cotton. It has an average tenacity of about 57.4 cN/tex (5.8g/dtex).

Elongation

Flax is a particularly inextensible fibre. It stretches only slightly as tension increases. The elongation at break is approximately 1·8 per cent dry, and 2·2 per cent wet.

Elastic Properties

Within its small degree of stretch, flax is an elastic fibre. It will tend to return to its original length when the tension is relaxed. It has a high degree of rigidity and resists bending.

Linen fabrics tend to crease, but this can be significantly reduced by modern crease-resisting treatments.

Specific Gravity. 1·54.

Effects of Moisture

Flax has a regain figure of about 12 per cent.

Linen is about 20 per cent stronger when wet than dry, which helps it to withstand mechanical treatment in laundering.

Effect of Heat

Highly resistant to decomposition up to about 120°C., when the fibre begins to discolour.

Effect of Sunlight

Gradual loss of strength on exposure.

Chemical Properties

Linen is more difficult to bleach than cotton, but modern methods of bleaching achieve whiteness with the minimum of chemical degradation.

Effect of Acids

Flax will withstand dilute, weak acids, but is attacked by hot dilute acids or cold concentrated acids.

Effect of Alkalis

Flax has a good resistance to alkaline solutions; linen fabrics can be washed repeatedly without deterioration.

Effect of Organic Solvents

Flax is not adversely affected by dry-cleaning solvents in common use.

Insects

Flax is not attacked by moth grubs or other insects.

Micro-organisms

When boiled and bleached, flax is virtually pure cellulose. Like other pure cellulose fibres, flax in this state has a high resistance to rotting. Under severe conditions of warmth, damp and contamination, however, mildews may attack the cellulose of flax, but resistance is generally high, particularly if the yarn or fabric is dry.

Other Properties

Flax is a good conductor of heat; this is one of the reasons why linen sheets feel so cool.

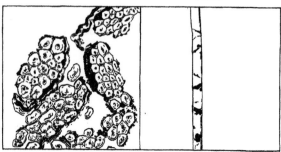

Flax

FLAX IN USE

In the past flax was in demand where extra strength and resistance to moisture were important. However, such flax products as sail and tent canvas, fishing lines and bookbinders' threads have now been replaced largely by synthetic substitutes. Leather-working thread, sewing thread and suture thread are still produced from flax. The fine household linen trade has declined greatly, but developments in blending with synthetics to give linen 'easy care' properties has ensured a long-term future for flax products. The use of union cloth (cotton and flax blended at the weaving stage) for furnishing fabrics is also established.

Waste flax fibre is made into high-grade banknote, writing and cigarette papers.

The ability of flax to absorb water rapidly is particularly useful in the towel trade. Linen glass-cloths will remove all traces of moisture from a glass without leaving any particles of fluff behind.

The molecular structure of the flax fibre makes linen an excellent conductor of heat; linen sheets are cool and linen garments are comfortable in hot weather.

Linen is often calendered or pounded in the roll by wooden hammers ('beetling') for as long as thirty-six hours. These treatments close up the fabric and bring out the beautiful finish that is characteristic of good linen.

Linen becomes stronger when it is wet, and will withstand repeated washings without deterioration. It is ideal for anything that has to put up with really hard wear.

The long life of linen fabrics was exemplified when Tut-ankh-amen's tomb was opened in 1922. Linen curtains, which had been there since about 1250 B.C., were still intact.

JUTE

In common with other bast fibres, jute has been used by man since prehistoric times. It comes from the inner bark of plants of the genus *Corchorus*, which probably originated in the Mediterranean area and was subsequently taken to India where it now grows profusely. Jute fabrics formed the 'sackcloth' of Biblical times.

The jute plant flourishes in hot, damp regions of Asia, and jute has for centuries been grown in enormous quantities for textile purposes. It is now produced in greater quantity than any textile fibre other than cotton. In 1976 – 77 some 3,468,000 tonnes of jute were produced, mainly in India (1,276,000 tonnes), Bangladesh (851,000 tonnes) and Thailand (183,000 tonnes).

During the latter half of the eighteenth century, the first shipments of jute reached Western Europe from India. In 1820, jute was spun experimentally at Abingdon near Oxford. The new fibre was of immediate interest to the flax and hemp spinners located at Dundee in Scotland. The Napoleonic Wars had cut off supplies of hemp and flax from Russia, and the Dundee mills began spinning jute in 1822. After ten years of experiment, the Dundee manufacturers were able to spin jute satisfactorily, and by 1850 the jute industry was well established. It was given further encouragement by the Crimean War which cut off hemp and flax supplies in 1853, and by the American Civil War of 1861–65 which interrupted the flow of cheap cotton.

Although other European countries took up the spinning and weaving of jute, Dundee has remained a centre of the industry. Meanwhile, India and Bangladesh have been steadily increasing the number of jute spinning and weaving mills, and both countries are now processing much of their own fibre.

PRODUCTION AND PROCESSING

The jute plant, *Corchorus*, is a herbaceous annual. It may grow to 5m (15 ft), with a stalk diameter of 20mm (¾ in). In India and Bangladesh, the plants are commonly harvested with a hand sickle.

Retting is carried out in a manner similar to that used for flax, the stalks being steeped in a sluggish stream of water. They are examined daily until the stage is reached at which the fibre can be separated easily from the stem. The strands of fibre, often as much as 2m (7 ft) long, are washed and hung up in the sun to dry. They are compressed into bales and sent off to the mills for spinning.

It is necessary to incorporate small amounts of mineral spindle oils into the fibre during conversion into yarn. Normal jute goods may contain up to 5 per cent oil, but so-called 'stainless' yarns containing 1 per cent of oil or less are commonly available when the jute·is to be used for special purposes, e.g. cables, fuses, carpet backings, wall-coverings, etc.

Bleaching and Dyeing

Jute is used very largely for cheap commodities such as sacks, bags and wrappings. Where necessary, and the extra cost is warranted, it is possible to bleach jute goods through various shades of pale cream up to pure white, and also to incorporate 'optical bleaches' (i.e. colourless dyestuffs which fluoresce a vivid white in daylight).

Dyestuffs of various types, as used for cotton, may also be applied to jute. The fibre has a special affinity for basic dyes, which provide brilliant effects even on unbleached base. Unfortunately, these effects are not very fast either to light or to water. Acid, direct and sulphur dyes are increasingly fast in this order, but also give increasing dullness of shade – all at reasonable cost. The increased demand for rugs, mats and carpets (especially cheaper tufted carpetings) has stimulated a corresponding demand for dyed jute yarns and fabrics suitable for these applications. Very bright and fast results are obtained with azoic and vat dyes, but their high cost limits their use with jute. The tendency for jute to turn brown in sunlight is a permanent disadvantage in better quality applications.

STRUCTURE AND PROPERTIES

Fine Structure and Appearance

Commercial jute varies from yellow to brown to dirty grey in colour, and it has a natural silky lustre. It consists of bundles of individual fibres held together by gummy materials, including the natural plastic lignin which plays an important role in the structure of all woody plants.

Jute usually feels coarse and rough to the touch, although the best qualities are smooth and soft. Retting destroys the cellular tissue that holds the bast bundles together, but does not normally separate the individual cells one from another. Some of the fibre-ends become detached from the strands, giving the jute its hairy, rough feel.

The individual cells of jute are about 2·6mm (1/10th in) long, on average. The cell-surface is smooth, but disfigured here and there by nodes and cross-markings. The fibres are coated with a layer of woody material.

Seen in cross-section, the cell is polygonal, usually with five or six sides. It has thick walls and a broad lumen of oval cross-section. By contrast with the regular lumen of flax, that of jute is irregular; it becomes narrow in places quite suddenly. Towards the ends of the cell, which are tapered, the lumen widens; the cell walls become correspondingly thin.

Jute contains about 20 per cent of lignin.

Tensile Strength

Jute is not so strong as flax or hemp, nor is it so durable. Individual fibres vary greatly in strength, owing to the irregularities in the thickness of cell walls.

Elongation

Jute fibres do not stretch to any appreciable extent. Jute has an elongation at break of about 1·7 per cent.

Elastic Properties

Jute tends to be a stiff fibre, owing to the part played by the material which cements the cells together.

Specific Gravity. 1·5.

Effects of Moisture

Jute is an unusually hygroscopic fibre. Its regain figure is 13·75 per cent. It can absorb as much as 23 per cent of water under humid conditions.

Effect of Age

If kept dry, jute will last indefinitely although the high content of non-cellulosic matter tends to make it sensitive to chemical and photochemical attack. Moisture encourages deterioration of jute, which loses strength with age.

Micro-organisms

Jute is more resistant to rot than either grey cotton or flax (i.e. uncleaned or unscoured). If lightly scoured it can have an excellent resistance owing to the protective effect of the lignin.

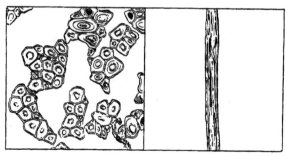

Jute

JUTE IN USE

Jute is cheap and reasonably strong, and is available in large quantities. These characteristics have enabled it to become an important fibre for sacks and packing cloths. These are used extensively for the storage and transport of agricultural products.

The resistance of jute fibres to stretching forces has proved a valuable property when jute is used for storage and transport purposes. Sacks and bales remain firmly in place after stacking; they do not distort and shift position as they would if made from a fibre more elastic than jute.

The hairiness of jute can be a disadvantage when jute sacks are used for food storage. The fibre-ends may break away and contaminate the food.

The finer qualities of jute are made into curtains and furnishing fabrics; mixed with wool, after treatment with caustic soda, jute is spun and woven into cheap clothing fabrics.

Familiar uses for jute include the following: Sacks, bags, baling and bundle cloths, wrappings (e.g. for bacon), bedding foundations, bonded fabrics, boot and shoe linings, mine brattice cloths and vent tubings, starched and glued buckrams and tailor's black packings, camp beds, cargo and other separation cloths (e.g. in rubber technology), cattle beddings, concrete cleavage fabrics, tarpaulins, damp courses, cables, plastics reinforcement, filter cloths, fire curtains, fuse yarns, furnishings, handbag and all types of stiff bag and case linings, hop pockets, horse covers, aprons of all heavy types, iron and steel tube and rod wrappings, canal linings (heavily bituminized), mail bags, motor car body linings, needlefelts, oakum, oven cloths, plasterers' scrim, prefabricated road and runways, roofing felt, rope soled sandals, trunk covering fabrics, tyre wrappings, upholstery foundations, strings for all purposes, certain ropings, wall coverings, wool packs, etc.

HEMP

In many parts of Asia, the fibre hemp has been in use since pre-historic times. Ancient records describe the use of hemp in China in 2800 B.C. During the early Christian era, production of hemp spread to the countries of Mediterranean Europe, and since then the fibre has come into widespread use throughout the world.

Like flax, hemp is a bast fibre. It comes from the plant *Cannabis sativa*, an annual of the family Moraceae, which grows to a height of 3m (10ft) or more.

The hemp plant is now cultivated in almost every European country, and in many parts of Asia.

Important producing countries include the Soviet Union, Yugoslavia, Roumania and Hungary.

PRODUCTION AND PROCESSING

The hemp plant is harvested and processed in a manner similar to that used for flax. Fibre is freed from woody matter by dew-retting or water-retting, followed by breaking and scutching. The fibre is softened by pounding it mechanically or by hand.

Hemp can be separated from the straw by a mechanical process more easily than in the case of flax. 'Green' hemp is now produced commercially in this way.

Dyeing

Hemp is used very largely in its natural state. When dyeing is necessary, direct colours are often used. Basic dyestuffs provide bright shades, the fibre being mordanted with antimony and tannin.

STRUCTURE AND PROPERTIES

Hemp is a coarser fibre than flax; it is dark in colour and difficult to bleach. The fibre is strong and durable, and is used very largely for making string, cord and rope.

Some Italian hemps are produced with great care; they are light in colour and have an attractive lustre similar to that of flax.

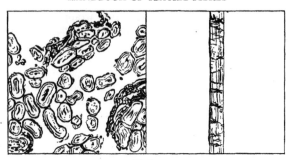

Hemp

Strands of hemp fibre may be 2m (6 ft) in length. The individual cells are, on average, 13 – 26mm (½ – 1 in) long. They are cylindrical in shape, with joints, cracks, swellings and other irregularities on the surface.

Like flax, the cells of hemp fibre are thick-walled; they are polygonal in cross-section. The central canal or lumen is broader than that of flax, however, and the ends of the cells are blunt.

The hemp fibre is more lignified than flax, and is consequently stiffer.

HEMP IN USE

During its long history, hemp has been used for almost every form of textile material. It has been made into fine fabrics by skilful spinning and weaving of carefully produced fibre, notably in Italy where a hemp fabric similar to linen is made. Nowadays, hemp is used mainly for coarse fabrics such as sacking and canvas, and for making ropes and twines.

Hemp can be 'cottonized' by a process similar to that used for flax, so that the individual fibres are freed. Cottonized hemp does not spin easily alone, but it gives useful yarns when mixed with cotton (up to 50 per cent hemp).

SUNN (Indian Hemp; Sann Hemp; Bombay Hemp)

The plant *Crotalaria juncea* has been grown in India and Pakistan since prehistoric times. It is a source of the bast fibre known as sunn; it was first brought to Europe during the early nineteenth century.

C. juncea grows to a height of 3m (10 ft), with stalks nearly 2.5cm (1 in) thick. It is cultivated and harvested in India and Bangladesh in a manner similar to jute. In 1976, the world production of sunn amounted to about 6,000 tonnes.

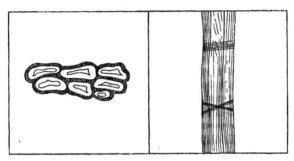

Sunn

Retting is carried out by steeping the stalks in water, and the fibre is peeled away from the rotted stalks by hand. It is dried in the sun and hackled (combed) by hand to remove any woody matter that may be adhering to it.

STRUCTURE AND PROPERTIES

Sunn is a light-coloured fibre when carefully prepared. It has a fine lustre.

Sunn fibre is almost as strong as hemp, and its strength is greater when wet.

Strands of commercial sunn are some 150cm (5 ft) long, and consist – like other bast fibres — of many individual fibres held together by natural gums.

The cells of sunn fibre are cylindrical, and are marked here and there by joints. The lumen is not so regular as it is in flax, and it is often filled with a yellow substance. Towards the ends of the fibre cells, the lumen disappears. The ends of the cell are blunt and rounded.

Seen in cross-section, the cells of sunn fibre are oval. There is a comparatively thick coating of lignin surrounding each cell.

Individual cells are about 8.5mm (⅓ in) long.

19

SUNN IN USE

Sunn fibre is used for cordage and paper manufacture. It is made into sacking and into carpet and rug materials.

The fibre has a good resistance to the effects of moisture, and is not readily attacked by micro-organisms present in sea water. These characteristics, allied to the increasing strength of sunn as it gets wet, have made the fibre particularly useful for fishing nets.

KENAF (Guinea Hemp; Mesta)

In Africa, there is a plant *Hibiscus cannabinus* which has long been used as a source of fibre for making cordage and coarse fabrics. The fibre is known as kenaf, Guinea hemp, or Mesta.

Kenaf has been grown in India for thousands of years, but the fibre was unknown in Western Europe until about two hundred years ago. It has been used since then as a sacking fibre, but did not really arouse any great interest until World War II. The shortage of jute during and after the war stimulated production of kenaf in the U.S., Cuba, Mexico and other countries.

Most of the world's kenaf is grown in India, Bangladesh and Thailand.

PRODUCTION AND PROCESSING

The kenaf plant is an annual, with 12mm (½ in) diameter stalks that reach 3m (10 ft) in height. It grows well in the hot damp climate of tropical countries.

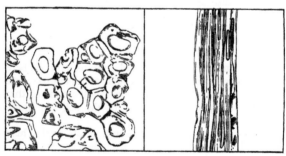

Kenaf

The methods used for harvesting and processing kenaf are similar to those used for jute. The stalks are retted and then beaten to free the fibre from unwanted material.

STRUCTURE AND PROPERTIES

Kenaf is a pale-coloured fibre which contains less non-cellulosic material than jute. It has the lustre which is characteristic of many bast fibres.

Kenaf has a breaking strength similar to that of low-grade jute, and it is weakened only slightly when wet.

The cells of kenaf are short, reaching only 6mm (¼ in) in length. They are cylindrical and the surface is striated and irregular. The lumen varies greatly in thickness at different points in the cell, sometimes disappearing altogether.

Seen in cross-section, the cell of kenaf fibre is polygonal and has a thick wall. It is coated with a layer of lignin. The fibre ends are thick and blunt.

KENAF IN USE

Most of the kenaf produced at present is used for making ropes and twines, and for coarse fabrics such as canvas and sacking. Some of the better quality fibre is made into carpet materials.

URENA

During World War II, the shortage of jute stimulated interest in another fibre, urena. This is a bast fibre that comes from the plant *Urena lobata*. It has been used from prehistoric times in Brazil.

The urena plant grows wild in many of the tropical regions of the world. It is cultivated as a fibre plant in the Congo and in Equatorial Africa. Brazil and Malagasy produce a considerable amount of urena, but most of it comes from plants growing wild.

Although urena has been used for centuries in those countries where the plant grows wild, it has been slow to develop as a commercially important fibre. The first attempts to cultivate urena were made in the Belgian Congo in 1926 and in 1929. During the 1930s, the crop became of increasing importance and cultivation was started in French Equatorial Africa. Since then other tropical countries have followed suit and many are now growing urena on a small scale. Most of it comes from the Congo Republic.

PRODUCTION AND PROCESSING

The urena plant is a perennial, with stalks that grow to a height of 3m (10 ft). After harvesting, the stalks are retted like jute in ponds or slow streams for 1 – 2 weeks. The fibres are stripped from the retted stalks by hand and washed to remove unwanted material.

STRUCTURE AND PROPERTIES

Urena has an attractive handle and appearance. It is near-white when carefully retted, and is soft to the touch. It has a natural lustre.

Strands of commercial urena are often 1.2m (4 ft) long. The fibre has a strength similar to that of jute.

Individual cells of urena are generally less than 2.6mm (1/10th) long. The lumen, as in the kenaf cell, is usually irregular.

URENA IN USE

Urena is made into yarns and fabrics similar to those that are made from jute. Much of it is used for sacking.

RAMIE (China Grass, Rhea)

Egyptian mummies of the pre-dynastic period (5000–3300 B.C.) were wrapped in fabric that has been identified as the bast fibre we now call ramie or China grass. This fibre comes from plants *Boehmeria nivea* or *Boehmeria tenacissema* which belong to a family of stingless nettles. The former is cultivated mainly in China and Formosa, and the latter in more tropical countries. *B. nivea* was grown as a source of fibre by the early Mediterranean civilizations, and during the Middle Ages came into widespread use in Europe. Ramie fibre was used by the early inhabitants of the American continent; it was made into twine for attaching the blades of knives and spears to handles and shafts used by American Indians.

During the eighteenth and nineteenth centuries, ramie cultivation became established in many areas of the Western world. Spinning mills were operated in England, France and Germany towards the end of the nineteenth century. But it is only in comparatively modern times that the production of ramie fabric has become established on a commercial scale.

In 1975 the world production of ramie was estimated at about 130,000 tonnes. The fibre is produced mainly in China (around 80% of total), South America, The Philippines, Korea, Japan and Indonesia.

The ramie plant is a perennial, sending up many stalks to a height of 1.2 – 2m (4 — 6 ft). The plants are hardy and grow well in warm climates. They are harvested when the lower stalks turn yellow and the new stalks are beginning to make their appearance.

Decortication

The ramie fibres are removed from the stalks by the process of decortication. This is usually carried out by hand; the process consists in peeling or beating the bark and bast material from the stalk soon after harvesting. The fibres are freed by soaking the bark in water and scraping with knives made from shells, bamboo, bronze or iron.

The long strands of ramie fibre are then dried and bleached in the sun.

The decortication process varies in detail in different regions. Sometimes the stalks are beaten against rocks before being peeled; the bark is battered with wooden mallets to free the fibre from adhering woody matter. In Indonesia the stalks are scraped in such a way as to leave the bast fibres clinging to the woody cores. These are then washed and the fibres are peeled away in the form of long ribbon-like strands.

During the 1930s great interest was aroused in the large-scale commercial possibilities of ramie as a textile fibre. But development of ramie production was held up by the primitive methods used in decortication, and attempts were made to devise machinery that could strip the fibres from the stalk.

Several machines are now in commercial use, and decortication has been mechanized.

Degumming

Before the ramie fibres can be spun, they must be released from the ribbons or strands in which they are held together by natural gums.

There are many degumming processes in use in different parts of the world. Where fibre production is carried out simply and by hand, the gums are removed by repeated soaking and scraping. Soda or lime may be used if these are available.

Commercial degumming is usually carried out by treating the fibres with caustic soda solution for as long as four hours. The fibres are then treated with bleaching powder, followed by immersion in a bath of dilute acid. The bleaching and acid-steeping are repeated until all the gum has been removed. Then the fibre is washed, oiled and dried.

Dyeing

Ramie can be dyed with all the classes of dyestuffs used for cotton, including direct, sulphur, basic, azoic and vat dyes. The techniques used are similar and the results are good. Dyeing is level and the fastness to light and washing is comparable with that of the same dyestuffs on cotton.

STRUCTURE AND PROPERTIES

Ramie fibre is white and lustrous. Ramie yarns may be as strong as flax line yarns. The fibre is durable but lacks elasticity.

Ramie absorbs water readily. Fabrics made from it will launder easily and dry quickly. They can be dyed readily.

Ramie yarn tends to have a hairy feel, due to the stiffness and coarseness of the fibres, which reduces their cohesion.

The cells of ramie fibre often reach 45cm (18 in) long. They are smooth and cylindrical, with thick walls. The surface of the cell is marked by little ridges.

The lumen narrows and disappears towards the ends of the ramie cell, which tapers to a rounded point.

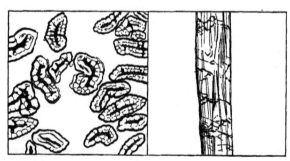

Ramie

RAMIE IN USE

Ramie is made into many types of heavy industrial fabric, such as canvas and packing materials. It is finding an increasing use in upholstery and furnishing fabrics and in clothes. Ramie yarns are used for fishing nets and sewing threads.

The lack of cohesion between ramie fibres and the consequent hairiness of ramie yarn makes it difficult to weave ramie into a smooth fabric. Much of the attractive natural lustre of ramie is lost when it is made into cloth.

This drawback can be overcome by a mercerizing process similar to that used for cotton. Ramie yarns are maintained under tension and treated with caustic soda. This brings about chemical and physical changes similar to those that take place in cotton.

NETTLE

Nettles have been cultivated for centuries in Scandinavia as a source of fibre for making sails. In other European countries, nettle fibres have long been used in small amounts for spinning and weaving into textiles. Certain species of stinging nettle are still grown in France and Germany for their fibre. The total output, however, is quite small.

PRODUCTION AND PROCESSING

Nettles are plants of the family *Urticaceae*. Of the thirty-odd species found in temperate climates, three are commonly used as sources of fibre: *Urtica dioica*, the Common or Great Nettle; *Urtica urens*, the small nettle; *Urtica pilulifera*, the Roman nettle.

U. dioica is a perennial; the other two are annuals.

After harvesting, the nettles are retted to free the bark from the woody core of the stalk. The bark is then boiled to release the fibres, which are hackled (combed) and oiled.

STRUCTURE AND PROPERTIES

Nettle fibres are creamy white to grey in colour, depending on the care taken in retting. They feel soft and pleasant to the touch. The strands are often about 1m (3 ft) in length.

The thickest fibres come from *U. dioica* which gives the highest yield. Fibres from *U. urens* and *U. pilulifera* are narrower, with thicker walls.

Individual fibre cells (*U. dioica*) may be 5cm (2 in). The surface of the cell is usually marked and distorted in many places. The lumen is narrow and contains a yellow material. The ends of the fibre cells are rounded; they are often broken into thin filaments.

Seen in cross-section the cell is oval, with thick walls.

Nettle

NETTLE FIBRES IN USE

Nettle fibres are made into twine and rope, and are woven into canvas and sailcloth. In some parts of Europe they are used for clothing and furnishing fabrics.

MISCELLANEOUS BAST FIBRES

There are many other bast fibres in use locally in different parts of the world. They come mostly from plants growing wild, and are used in the main because they are cheap and ready to hand.

THE LEAF FIBRES

The leaves of monocotyledenous plants are held in shape and strengthened by fibres which run in hawser-like strands through the length of the leaf. These leaf fibres are often of great commercial value, and are used in large quantities for making ropes and cordage, and for the production of textile fabrics.

In general, leaf fibres are coarser than the fibres which come from the bast of dicotyledenous plants. Bast fibres are commonly described as 'soft' fibres and the leaf fibres as 'hard' fibres. This classification is not, however, a rigid one; some leaf fibres are softer than some bast fibres.

SISAL

The ancient Mexicans and Aztecs clothed themselves in fabric woven from the fibre known as sisal. This is a leaf-fibre that comes from the plant *Agave sisalana*, which is indigenous to Central America. It derives its name from the Yucatan port of Sisal on the Gulf of Mexico.

The sisal plant is now cultivated widely in East Africa, Mexico, Haiti, Brazil and in other regions of South America. The world output (1978) is in the region of 550,000 tonnes.

PRODUCTION AND PROCESSING

Sisal plants send up huge leaves almost from ground level. The leaves are firm and fleshy, and form a rosette on a short trunk.

After six or seven years of growth, the sisal plant sends out a flower stalk that rises to some 6m (20 ft). When it has flowered, the plant produces tiny buds which develop into small plants. These fall to the ground and take root, and the parent plant dies.

Leaves are harvested when the plants are 2½ to 4 years old and at intervals until the plant eventually dies. A good plant may yield 400 leaves during its lifetime, and each leaf may contain up to 1,000 fibres. The outer mature leaves are cut away and treated in machines which scrape the pulpy material from the fibres. After washing, the fibre is dried and bleached in the sun, or oven-dried.

Dyeing

Sisal has a good affinity for direct cotton and acid dyestuffs, which provide attractive shades of good light fastness.

Direct dyestuffs are used in the same way as in the dyeing of cotton. Acid dyes are applied from a neutral or acid dyebath.

Basic dyes are commonly used for dyeing sisal which is used in ropes. They have poor light fastness and are less satisfactory than direct or acid dyes when the sisal is used for matting.

STRUCTURE AND PROPERTIES

Strands of commercial sisal are 60 – 120cm (2 – 4 ft) in length. They are strong and consist of many individual fibres held together by natural gums. If processing has been carried out carefully, the sisal is creamy-white in colour.

Sisal fibre tends to be stiff and rather inflexible. It absorbs moisture readily and is weakened by being steeped for long periods in salt water.

There are a number of different types of cell in a typical specimen of sisal. The 'normal' fibre cells are straight and stiff; they are cylindrical and often striated. The average length is about 2.5mm (1/10th in). These fibres sometimes appear saw-edged and have tapering ends. The lumen varies in thickness and definition; the cell walls are thick where the lumen is thin and vice versa. The lumen is often packed with tiny granules.

Sisal also contains broader fibres with a characteristic lattice pattern and with small pore-markings. Some cells are cushion-shaped and others are short and rectangular. Here and there, small spiral-shaped bodies can be seen, like little springs.

Sisal contains about 6 per cent of lignin (based on dry material).

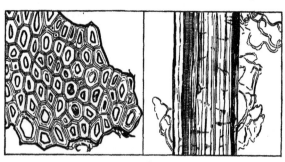

Sisal

SISAL IN USE

Sisal is one of the most valuable of all cordage fibres. It is too stiff to be used satisfactorily for certain purposes, such as power trans-mission, in which it has to run through pulleys or over wheels. Prior to World War II, sisal ropes were also regarded as being of limited use in marine cordage; it was believed that sisal deteriorated too rapidly in salt water. Experience during the war showed, however, that this is not the case, and sisal is now widely used for marine ropes and hawsers particularly in under-developed areas.

Sisal is used extensively for making baler and binder twine, and for sacks, paper filters and other industrial uses.

The high strength, lustre and good colour of sisal have made it into an attractive fibre for certain textile uses. It is made into matting and rugs. Its ability to take up direct cotton and acid dye-stuffs has made it a popular fibre for ladies' hats.

HENEQUEN

In Yucatan, the fibre henequen is produced from a plant *Agave fourcroydes*, which is closely related to the *A. sisalana* which yields sisal. Henequen fibre is also a valuable export from Cuba. It is often known as 'Yucatan sisal' or 'Cuban sisal'.

Henequen is an important leaf-fibre; the total output is in the region of 150,000 tons a year. Mexico produces about four-fifths of the world's henequen, most of which comes from the State of Yucatan. The remaining fifth comes almost entirely from Cuba.

PRODUCTION AND PROCESSING

The henequen plant is very similar to that from which sisal is obtained. The leaves are prickly and grey-green in colour. They are first cut, one or two at a time, from plants about six or seven years old. Then, for fifteen to eighteen years, a few leaves are removed twice yearly until the plant flowers and dies.

The leaves are processed in the same way as sisal and the strands of washed fibre are dried in the sun.

STRUCTURE AND PROPERTIES

Henequen is very similar to sisal. The strands of fibre, 150cm (5 ft) long, are of good colour and have an attractive lustre. They are usually finer than strands of sisal fibre.

The individual cells of henequen are almost identical with those of sisal.

HENEQUEN IN USE

Henequen is used in much the same way as sisal. It provides much of the world's agricultural twine.

Coarse fabrics such as canvas have been made from henequen in Mexico since prehistoric times.

ABACA (MANILA)

The *Musaceae* family of plants is one of the most useful in the world. It provides us with all manner of foods and industrial raw materials. *Musa sapientum*, for example, gives us the banana; *Musa textilis* is a source of the paper-making and cordage fibre abaca, or Manila hemp.

The abaca plant is indigenous to the Philippine Islands; native islanders were making textiles from its fibres when Magellan visited the islands in 1521 during his circumnavigation of the globe.

During the early nineteenth century, supplies of abaca began to reach the Western world, and its value as a cordage fibre was quickly appreciated. It was better than hemp for many purposes, particularly in marine ropes and hawsers.

Despite the many attempts that have been made to establish abaca production in other parts of the world, the Philippine Islands remain the chief source of the fibre. Total production in 1977 was 75,000 tonnes, of whch some 85 per cent came from the Philippines. The remainder came from Ecuador.

PRODUCTION AND PROCESSING

M. textilis grows easily in the Philippines and needs little cultivation. The plant comprises a cluster of sheath-like leaf stalks. The stalk is composed of a fibreless pulpy centre core surrounded by overlapping leaf-sheaths. Each sheath contains a thin layer of fibre.

These stalks often reach a height of 7.5m (25 ft). After one and a half to two and a half years' growth the blossom appears on some of these stalks, normally on two to four. This is the most satisfactory stage for fibre production, and these stalks are then cut down near the ground level. By this time the plant consists of from ten to thirty stalks in various stages of growth, and two to four of these reach maturity in four to six months after the previous cutting. The diameter of mature stalks is usually 13 – 30cm (5 – 12 in). The average useful life of the plant is about fifteen years, although some varieties continue producing for up to thirty years.

Fibre Extraction

The fibre is extracted by separating the ribbons of fibre from the layers of pulp. These ribbons, which are known as tuxies, are then drawn under a knife, usually made of metal, and the residual pulp is removed from the fibre, which is then hung up to dry.

Grades

The leaf sheaths vary in colour and texture according to their position on the stalk. The sheaths comprising the stalk normally fall into four groups. The outside sheaths (baba) are of dark brown or light purple and green strips. (This discoloration is due to exposure to the sun.) The sheaths next to the outside (segunda baba) are striped very light green and purple. The middle sheaths are a very light green or light yellow. The inner sheaths (ubud) are almost white. The outer sheaths produce the strongest and the inner sheaths the weakest fibre. The tuxies from these four groups of sheaths produce basically four qualities of fibre. In practice many qualities are produced by varying the type and pressure of the knife used for removing the residual pulp.

Other factors affecting the quality of the fibre are the condition of the stalks; immature or over-mature stalks reduce the quality as also does delay in removing the tuxies from a stalk after it has been cut, and delay in drying the fibre after extraction.

STRUCTURE AND PROPERTIES

Commercial abaca fibre is in the form of strands containing many individual fibres held together by natural gums. The strand-length varies greatly depending on the precise source and treatment of the fibre during processing. Good quality abaca is often in the form of strands up to 4.5m (15 ft) long.

Abaca has good natural lustre. Its colour depends upon the conditions under which it has been processed; good quality abaca is off-white, whereas some poor quality fibre is nearly black.

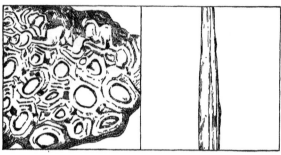

Abaca

Abaca is strong and sufficiently flexible to provide a degree of 'give' when used in rope. The fibre is not readily affected by salt water. It has a slight natural acidity which can cause corrosion when abaca is used as a core in wire ropes.

Individual fibre cells are cylindrical and smooth-surfaced. They are as much as 6mm (¼ in) long, are are regular in width. The ends taper gradually to a point.

In cross-section, the fibres are polygonal and the cell walls thin. The lumen is large and distinct; it is round and uniform in diameter although both fibre and lumen show occasional constrictions. In places, the lumen contains granular bodies.

Abaca fibres are largely cellulose (about 77 per cent of moisture-free fibre), but are coated with considerable amounts of lignin (about 9 per cent). The individual fibres can be freed by boiling the strands in alkali.

Abaca which has been treated in this way contains epidermal cells which are almost rectangular. They cling together forming little rafts among the fibrous cells.

ABACA IN USE

Most of the abaca produced today is used in the manufacture of strong high-grade paper, such as tea-bags, stencil tissue, meat casing and disposables. Some is still used for ropes and cordage. The fibre's resistance to the effects of sea-water, and its natural buoyancy, have created a ready market for it in the manufacture of hawsers and ships' cables.

Abaca fibre is also used for making hoisting and power-transmission ropes, well-drilling cables, fishing nets and lines, and other types of cordage where strength, durability and flexibility are essential.

Some of the fine inner fibres from the abaca leaf-stalk are used directly, without spinning, for making delicate, lightweight, yet strong fabrics. These fabrics are used in the Philippines for clothing, and for hats and shoes. Some abaca is used for carpets, table mats, etc.

OTHER LEAF FIBRES

1. Canton
SOURCE: Philippines, from plant of *Musa* species similar to *M. textilis* (abaca).
CHARACTERISTICS: Similar to abaca.
USES: Cordage.

2. Pacol
SOURCE: Philippines, from plant of *Musa* species.
CHARACTERISTICS: Weaker and softer than abaca.
USES: Cordage.

3. Cantala (Maguey)
SOURCE: *Agave cantala*. Native of Mexico; grown in Philippines, India, Indonesia.
CHARACTERISTICS: Similar to henequen. Softer than henequen and more supple. Fibre strands are of round cross-section (c.f. sisal – horseshoe shape). Individual fibres long and fine. Thicker cell walls and narrower lumens than sisal.
USES: Same as sisal and henequen. Native fabrics.

4. Letona
SOURCE: *Agave letonae*. Grown commercially in El Salvador.
CHARACTERISTICS: Similar to henequen. Lustrous strands.
USES: Coffee bags, canvas, hammock cloth.

5. Mauritius fibre
SOURCE: *Furcraea gigantea*. Resembles Agaves in growth and habits. Island of Mauritius.
CHARACTERISTICS: Long white, lustrous strands. Similar to sisal and henequen, but weaker, finer and softer.
USES: Bagging and coarse fabrics.

6. Phormium (New Zealand 'flax' or 'hemp')
SOURCE: *Phormium tenax*. Indigenous to New Zealand. Leaves about 3m (10 ft). Fine native fabrics noted by Captain Cook. Also grown commercially in St Helena, the Azores, South Africa and in some South American countries.
CHARACTERISTICS: Fairly strong; flexible, good lustre, good resistance to sea-water. Softer and weaker than Manila hemp. Individual fibres smooth with pointed ends, average length about 6 mm. Cross-section round with thick cell walls. Narrow circular lumen tapering and disappearing towards ends.
USES: Ropes, twines, coarse bagging materials.

7. Sansevieria

SOURCE: Various species of genus *Sansevieria*. Indigenous to Africa and India. Cultivated in Yucatan, Mexico.

CHARACTERISTICS: White and soft to touch. Good lustre. Fairly strong and resilient. Good resistance to sea water.

USES: Ropes and twines. Bagging fabrics.

8. Caroa

SOURCE: *Neoglazovia variegata*, indigenous to Central and South America. Grows wild in Brazil in large quantities.

CHARACTERISTICS: Creamy white. Strong, flexible and fairly soft to touch. High proportion of lignin.

USES: Ropes and twines. Clothing fabrics, e.g. for lightweight suits.

9. Pineapple fibre

SOURCE: *Ananas comosus*, the pineapple plant. Cultivated in many parts of world, principally Hawaii, Philippines, Indonesia, India, West Indies.

CHARACTERISTICS: Soft, white and good lustre. Strong and hard wearing. Smooth-surfaced cells, tapering to pointed ends. Narrow lumen. Oval cross-section.

USES: Cordage and threads. Woven into fine cloth of great beauty.

10. Pita floja

SOURCE: *Aechme magdalenae*. Central and South America. Grows wild in Brazil.

CHARACTERISTICS: Creamy white. Good lustre. Strong, Good sea-water resistance.

USES: Ropes, twines, coarse fabrics.

11. Bromelia

SOURCE: Species of *Bromelia*. Indigenous to Central and South America. Cultivated in Mexico for fibre.

CHARACTERISTICS: Fine, soft, lustrous fibre. Creamy white.

USES: Canvas-type fabrics, ropes and twines.

12. Palma

SOURCE: *Samuela carnerosana*. Native to Mexico.

CHARACTERISTICS: Strong.

USES: Ropes and twines, bagging fabrics, brushes.

13. Fique (Cabuya)

SOURCE: *Furcraea macrophylla*. Colombia. Wild and cultivated.

CHARACTERISTICS: Coarse fibre.

USES: Bagging fabrics, coffee bags, canvas for hammocks, etc.

14. Piassava

SOURCE: Leaf stalks of palms.
Bahia piassava (*Attalia funifera*).
Para piassava (*Leopoldinia piassaba*) exported from Brazil.
Stalks retted.

CHARACTERISTICS: Stiff fibres.

USES: Brooms and brushes.

34

15. Raffia
 SOURCE: Madagascar. Strips of fibrous material from surface of leaflets of certain palms.
 CHARACTERISTICS: 1.2 – 1.5m (4 – 5ft) strands. Pale creamy colour.
 USES: Horticulture, hats, etc.

THE SEED AND FRUIT FIBRES

The seeds and fruits of plants are often attached to hairs or fibres which, like other plant fibres, are constructed in the main from cellulose. Many of these fibres are used in the textile industry; one of them – cotton – has become the most important textile fibre in the world.

COTTON

Though the bast and leaf fibres are of very great value to the world, they cannot begin to compare in importance as textile fibres with the seed fibre, cotton.

Cotton is the backbone of the world's textile trade. Many of our everyday textile fabrics are made from cotton; fabrics that are hard-wearing and capable of infinite variety of weave and colouring.

Like the other plant fibres, cotton is essentially cellulose. But it is not produced by the plant as part of its skeleton structure, as are the bast and leaf fibres. Cotton is attached to the seeds of plants of the *Mallow* family; the fibre serves probably to accumulate moisture for germination of the seed.

Early History

The idea of using these fine seed-hairs as textile fibres came at an early stage in textile manufacture. Cotton fabrics were made by the Ancient Egyptians and by the earliest of Chinese civilizations. Samples of cotton materials have been found in Indian tombs dating back to the year 3000 B.C. There is some evidence that cotton may have been in use in Egypt in 12,000 B.C., before the use of flax was known. Specimens of woven cotton fabric have been found in the desert tombs discovered in Peru. These pre-Inca textiles were designed and woven with immense skill. They include brocades and tapestries, crocheting and lace.

No matter where the spinning and weaving of cotton may have been developed first, there is no doubt that India was the true cradle of the cotton industry. Cotton fabrics of remarkable quality were being produced as early as 1500 B.C., using only the most primitive of spinning and weaving techniques.

As textile skills developed in India, many different types of fabric were produced. Brocades and heavy fabrics, embroidered materials and muslins were made in great variety and of incredibly high quality. By the seventeenth century, Indian textile craftsmen were spinning and weaving cotton into the fabulous Dacca muslin; this beautiful fabric was so light that 66m (73 yd), 90cm (1 yd) wide, weighed only 454g (1 lb) – less than one-quarter of the weight of a modern fine-quality muslin. These Dacca muslins were used for royal and ceremonial occasions. They were woven from yarn spun entirely by hand, using only a simple spinning stick.

At the time of the Roman Empire, cotton growing and manu-facture became established around the shores of the Mediterranean. Trade with India developed, and lasted until the Roman Empire collapsed, bringing with it a breakdown in trading activities between the Mediterranean and the East.

During the seventh century A.D. the Saracens built up their Mediterranean empire and reached to the borders of India itself. Once again trade grew between India and the Levant. New caravan routes were established and commerce thrived as it had never done before. The foundation of much of this intercontinental trade was cotton.

Cotton was being grown on the Greek mainland from the eighth century A.D., and its cultivation became established in other European countries. As the Moors penetrated into Spain, they carried their technical and artistic knowledge with them. Some of the finest fabrics were made in Spain during the tenth century; Cordova, Seville and Granada were centres of the cotton weaving and dyeing trades, and their products compared favourably with those of Eastern cities.

Europe, however, was too involved in religious struggles to allow of any volume of trade with Spain. When the Moors were eventually driven out their skills in cotton cultivation and manufacture went with them.

During the twelfth and thirteenth centuries, the Crusades brought Europe into contact with the arts and crafts of Eastern countries. Once again, cotton became one of the most important of the articles that flowed through these great trade routes, and the textile industries of southern France and northern Italy began to flourish. The trading centres of Genoa and Venice became the gateways of the European continent.

Cotton in Europe

One of the earliest mentions of cotton in European history records the weighing of cotton in the public scales at Genoa in 1140. This cotton came from Antioch, Alexandria and Sicily. Venice, however, is believed to have been the first European city (outside Spain) to manufacture cotton fabrics. Here, cotton from the Mediterranean was spun and woven into cloth for Europe, and during the next three hundred years the textile trade flourished along the eastern Mediterranean shores. Cotton became an established article of commerce between Venice and cities of Central Europe. During the Middle Ages, Germany became an important centre of European cotton manufacture.

Despite this steady development of cotton spinning and weaving in parts of Europe, the finest calicos and prints were still brought by caravan from India. Europe had not yet been able to match the skill of Asian craftsmen.

In 1368, the great Mogul Empire in Central Asia, which had favoured commerce with the West, was succeeded by the Ming Dynasty. The overland routes were closed and as the Turks conquered Syria and Egypt the trade with India virtually ceased. This was responsible more than anything else for the great voyages of exploration that made the fifteenth century live in history. The search for a water route to the East led Columbus to America and took Vasco da Gama on his voyage to Calcutta via the Cape of Good Hope.

For the next hundred years, Portugal monopolized the sea-borne cotton trade that grew between Europe and the East. From Lisbon, cotton was shipped by the Dutch to Antwerp, Bruges and Haarlem, which became the textile centres of eastern Europe.

As Spain's sea power grew, trade between British and Dutch ports and the East was diminished, but the defeat of the Armada in 1588 opened up once and for all the trade routes from Europe to India. This was the time of the great trading companies that were built up largely on the cotton trade. The British East India Company formed in 1600 was followed within a few years by those of the Dutch and French; the first steps had been taken towards the creation of the European empires. And cotton was a commodity with which these trading companies were very much concerned.

Meanwhile, the manufacture of textiles was growing into an industry in Britain. Thousands of Protestant artisans had fled from religious persecution on the Continent, bringing with them their skill in the textile crafts. By the end of the seventeenth century, Britain had become an exporter of cotton fabrics – a trade in which she was ultimately to establish the leadership of the world.

The Industrial Revolution

The eighteenth century saw textiles developing as a national industry in place of the local guilds that had served during mediaeval times. In 1736, the Manchester Act wiped out the law of 1700 which had been engineered by the wool merchants to prohibit the sale of cotton goods in England. Lancashire then set out to become the cotton centre of the world.

During the latter half of the eighteenth century, the Industrial Revolution in Britain transformed production methods in the textile industry. This was the age of invention that was to make Britain the workshop of the world.

At this time, most British cotton was being imported from the West Indies, with a certain amount from India, the Levant and Brazil. Cotton has been planted and grown for commercial use in Virginia since 1607, but the production of American cotton was held up owing to difficulty in removing the cotton from the seeds to which the fibres clung tenaciously. In 1793, Eli Whitney invented the cotton gin, a machine for removing the fibres from cotton seeds, and America entered the world market as a cotton producer. By 1800, she was exporting 8 million kg (18 million lb) a year to England; eleven years later this had reached 28 million kg (62 million lb).

From this time on, Lancashire was to reign unchallenged as the cotton manufacturing centre of the world – a position she held until the First World War.

Since then, the great consumer countries of the East have developed their own manufacturing industries, and Lancashire no longer holds her former position in the textile world. Many factors have combined to intensify the changing conditions in the cotton trade; better home heating and changing fashions have reduced the clothing worn by women to a quarter of the yardage carried by their Victorian grandparents. And man-made fibres like nylon and rayon are encroaching steadily in the fields where cotton was previously employed.

In 1964 – 65 the world production of cotton amounted to 11,300 million kg The most important producing countries are shown in the table.

The Cotton Boll

Formation of the Fibre

Cotton grows inside the seed pods of a wide variety of plant species included in the *Gossypium* family. The early primitive cottons grew naturally as perennials, and for many years cultivated cotton was also grown as a perennial. In the tropics, perennial cotton plants may grow 6m (20 ft) high. Nowadays, with only one or two major exceptions, the world's cotton is grown by raising annual crops, the plants reaching a height of between 1.2 and 1.8m (4 – 6 ft).

Cotton seed is usually sown in the spring and the young plants are thinned out later into rows. In due course, many creamy-white flowers appear, which turn pink towards the end of the first day. On the third day, the flower withers and dies to leave a small green seed pod, or 'boll'.

The cotton fibres form on the plant as long hairs attached to the seeds inside the boll. As the plant grows, the fibres are packed tightly into the boll. When it reaches maturity, the boll bursts and the cotton appears as a soft wad of fine fibres.

The individual cotton fibre is a seed-hair consisting of a single cell. It grows from the epidermis or outer skin of the cotton seed.

Each cotton seed may produce as many as 20,000 fibres on its surface, and a single boll will contain 150,000 fibres or more. The boll itself is a fruit which forms when the flowers drop from the

39

Production of Cotton during Season
1979 *thousand tonnes)*

U.S.A.	3,163
U.S.S.R.	2,821
China	2,207
India	1,220
Egypt	482
Mexico	336
Brazil	575
Pakistan	650
Turkey	505
Peru	72
Argentina	140
Sudan	131
Syria	132
Iran	110
Spain	40
Uganda	13
Colombia	108
Greece	100
Nigeria	43
Salvador	72
Mozambique	15
Nicaragua	109
Tanzania	60
Eastern Europe	9
Guatemala	146
Afghanistan	38
Burma	17
Paraguay	71
Korea	6
Angola	13
South Africa	47
Zimbabwe, Malawi, Zambia	45
Kenya	11
Australia	53
West Indies	1
Israel	75
TOTAL	14,050

cotton plant. The young fruit that remains increases in size for perhaps seven weeks, forming the ripe boll; this then opens to expose the mass of cotton fibres which expand and dry into a light fluffy mass.

The growth of these cotton fibres takes place throughout the boll-ripening period. In some varieties of cotton, tiny fibres can be detected on the embryo seed one or two days before flowering. Other varieties begin their fibre-production a day or two after flowering.

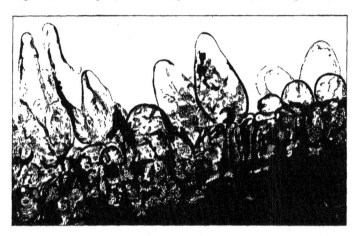

Cotton. The surface of a very young seed, showing infant cotton fibres forming – *After Mary L. Rollins, U.S. Dept. of Agriculture.*

Fibre Growth

During the first week after the cotton plant has flowered, hundreds of fibres appear from the seed coat. For several days, more and more young fibres continue to thrust their way out of the seed until each seed is carrying a 'crop' of thousands of individual fibres.

For six days, the growth of the young cotton fibre is comparatively slow. Then for the next fifteen days it is much more rapid; the fibre may reach a length equal to 2,000 times its diameter during this three-week growing period. Then for three days it grows more slowly again until the lengthwise growth comes to a sudden stop.*

During its period of rapid elongation, the cotton fibre is in the

* The time-scale given in this section represents that of a typical cotton plant; it varies considerably according to variety and conditions of growth.

form of a thin-walled tube of cellulose with one end closed and the other attached to the seed. It is filled with protoplasm and liquid nutrients which have been drawn from the main supply vessels of the plant. It resembles a long thin balloon distended with water from a tap. Magnified to the thickness of a finger a typical fibre would be about 15m (50 ft) long.

Growth Rings

When it stops its lengthwise growth, the cotton fibre begins to strengthen its internal structure. Layers of cellulose are added one after another to the thin cellulose membrane from inside the cell. Each day sees a new layer deposited, creating a structure similar in cross-section to the growth rings in a tree. The cotton fibre, however, adds its layers by depositing cellulose from the liquid inside the fibre; the innermost layers are the youngest ones, whereas the outermost layers are the youngest in a tree.

Each growth-ring in the cotton fibre corresponds to a day of growth and cellulose-deposition. Every ring consists, in fact, of two layers, one solid and compact and the other porous.

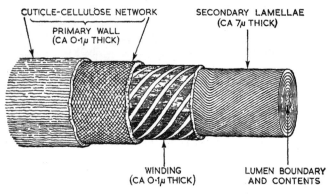

Cotton. A diagram showing the layered components of a cotton fibre cell wall – *After Mary L. Rollins, U.S. Dept. of Agriculture.*

Experiments have shown that cotton grown under constant artificial illumination and constant temperature has no growth-rings. On the other hand, cotton grown in artificial light switched off and on develops rings like the natural fibre. The cellulose is laid down in the form of spiral fibrils or tiny threads, some 1,000 or more to each ring. The deposition of cellulose continues for about twenty-four days, so that each mature cotton fibre can be regarded as being constructed from thousands of fibrils of cellulose arranged in spiral form.

When the boll opens, the moisture evaporates from the fibres. Until this happens, the fibres maintain their tube-like appearance, with a circular cross-section. But as the fibres dry out, the cell walls collapse, forming a ribbon-like structure that resembles a bicycle inner-tube from which the air has been removed.

During its period of growth, cotton is compressed tightly into the limited accommodation available inside the boll. As the cell walls thicken, the fibres are fixed in their distorted positions. Then when the boll bursts and the fibres dry out in the air, they twist lengthwise, forming convolutions which are characteristic of the fibre. These twists take place in both directions in the fibre; some are left-handed and others right-handed, with an almost equal number of each in any individual fibre.

The number of convolutions varies greatly; on average, a fibre will have some 50 twists per cm (125 twists per in).

Effect of Growth Conditions

The basic characteristics of the cotton fibre, such as its diameter, are determined by hereditary factors. But the nature of the fibre is also affected greatly by the conditions under which the plant is grown.

In a typical plant, the cotton will be developing for perhaps seven weeks inside the boll. But all the bolls do not mature at the same time; bolls may go on opening on a cotton plant for a period of eight or nine weeks. The total time between flowering and the opening of late bolls may thus amount to about four months.

The ultimate yield of cotton is affected by the conditions under which the plant has grown before it flowered. The quality of the fibre is influenced by the conditions it experiences during the time that the fibres are developing inside the boll.

A setback in the growth of the plant during the period of fibre development may slow up the deposition of cellulose. When conditions are restored to normal, the fibre-growth will continue, but the effects of the interruption will be shown in the quality of the fibre. Short fibres result from poor growth in the initial phase of development; thin-walled fibres arise from interrupted or slowed-down development of the wall in the second phase.

43

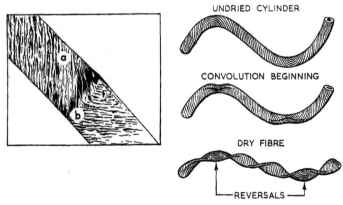

UNDRIED CYLINDER

CONVOLUTION BEGINNING

DRY FIBRE

REVERSALS

Cotton Fibre: Convolutions. As the cotton fibre grows, layers of cellulose are laid down inside the thin primary wall. Each layer corresponds to a day's growth, and the cellulose is laid down as fibrils which form a spiral pattern round the long axis of the fibre (a). The layers of cellulose laid down in this way form the secondary wall which makes up the major part of the fibre.

The growing fibre is a tiny tube of near circular cross section. The lumen acts as a channel through which materials are carried along the fibre.

When the boll opens, the fibres dry and collapse into flattened tubes. Convolutions appear, the bends in the fibre corresponding to places in the fibre where the spiral pattern of fibrils reverses its direction (b) – *After U.S. Dept. of Agriculture.*

No matter how the cotton is grown, it is inevitable that many of the fibres in every boll will be in an immature state. The proportion of immature fibre to mature fibre is an important factor in determining the quality of the cotton.

In ordinary commercial cotton, about one-quarter of the fibres will be immature. Sometimes, the proportion of mature cotton reaches 90 per cent, but such high 'maturity counts' are rare. In commercial upland cotton, maturity counts of more than 84 per cent are described as 'hard-bodied'. Average maturities lie between 68 and 76 per cent and cottons with maturity counts below about 67 per cent are regarded as immature ('soft-bodied' or 'weak').

Neps

The immature, thin walled fibres can cause trouble in several ways. They do not dye to as dark a shade as the thick walled fibres; they break easily, causing greater losses of waste fibre during processing. They are more flexible than the thick-walled fibres, so that they bend and tangle more easily, forming 'neps'. If these 'neps' appear in the dyed cloth they show up as specks of lighter shade.

Production of Cotton

Cotton is usually picked by hand or by machine in the autumn. As the bolls on a plant do not all mature at the same time, a field has to be picked over several times before the crop is in. Mechanical pickers have been developed during recent years, and are in almost universal use. The cotton they pick includes a high proportion of leaves and bolls, which increase the amount of cleaning needed and also tend to stain the fibres. In the U.S., more than 99 per cent of the crop is now mechanically harvested. Mill machinery has been devised to improve the cleaning of mechanically picked cotton.

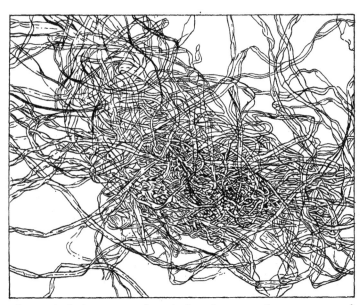

Cotton. A single nep seen through the microscope, showing a mass of tangled, immature fibres – *After Mary L. Rollins, U.S. Dept. of Agriculture.*

Boll Weevil

Cotton growing is now one of the major agricultural industries of the world. Yet cotton is subject to attack from a wider range of insects and diseases than almost any other plant. In particular, a small beetle called the cotton boll weevil (*Anthonomus grandis*) is responsible for tremendous losses in the cotton crop every year. This beetle is a native of Mexico that first appeared in the cotton fields of Texas towards the end of the last century.

Crossing the Rio Grande river in 1892 it spread rapidly eastwards and northwards. By 1903, the boll weevil had reached Louisiana; four years later it was in Mississippi. In 1909 the insect was in Alabama and by 1915 it was attacking cotton in Georgia. By 1921, the boll weevil had spread over the entire cotton-growing area of the United States, causing damage to the crop amounting in bad years to as much as 200 million dollars.

The boll weevil is a tiny beetle that feeds on the buds and bolls of the cotton plant. It has a long snout with which it is able to bite into the plant; it then feeds on the interior of the plant and lays its eggs inside the puncture. Three days later the eggs hatch into fat white maggots which begin feeding on the boll. After a week to twelve days they are fully grown and turn into pupae from which the weevils emerge after three to five days. Then in another week's time the weevil is ready to start rearing a family on its own account. The whole life cycle takes only three to four weeks; barring losses the offspring from a single pair of weevils could amount to several millions during a season.

Cotton growers today have achieved a high degree of control over the boll weevil. Quick-maturing varieties of cotton are grown when possible to avoid the worst season for attack; cultivation methods have been improved and modern synthetic insecticides have carried the war to the insect itself.

Other Pests and Diseases

The pink boll worm (*Pectinophora gossypiella*) is another insect pest that causes great economic loss in all cotton growing countries, especially in India, Egypt and Brazil. This insect has damaged the Egyptian cotton crop to the extent of £8 million or more in a single season.

The adult insect is a small brown moth which leaves its eggs on the cotton plant. Larvae hatch from the eggs and tunnel into the bolls. They eat the seeds of the cotton, preventing proper development of the fibre. Cotton attacked by the boll worm is often stained pink.

Carbon disulphide is used to destroy the pink bollworm; in Egypt, cotton seed is heat treated to prevent infestation.

Other insect pests, such as the cotton leaf worm, cotton boll worm, cotton stainer and cotton red spider, attack the cotton crop in different countries, causing great damage if adequate precautions are not taken.

The cotton plant, like all other living things, is subject to attack by disease-producing micro-organisms. Cotton wilt and cotton anthracnose are fungus diseases; Angular leaf-spot or Blackarm is caused by bacteria; root knot is caused by a tiny worm that enters the roots of the cotton plant from the soil. Black rust is a deficiency disease that is common in regions where the soil is lacking supplies of essential nutrients.

A World Crop

Cotton cultivation has now become a major industry in sixty countries and the plant has adapted itself to a range of climatic conditions. Its essential needs are a growing period of up to six months with plenty of moisture and sunshine, and a dry period that enables the plant to mature. Such conditions are found in general between the north and south latitudes of 40 degrees.

The U.S., U.S.S.R., India, China, Mexico, Brazil, Egypt, Pakistan, Turkey, Argentina and Peru are main centres of cotton cultivation; the U.S. produces almost one fifth the total crop

Altogether, over 10 million tons of cotton are grown every year. The United States cotton belt alone has more than a million farms with cotton as the main crop; the average yield per acre rose from 47.7kg (105 lb) in 1925 to 197.5kg (435 lb) in 1957.

Types of Cotton

Many varieties of cotton plant are grown commercially in different parts of the world, under a wide range of growing conditions. As a result, there are a great many different grades and qualities of cotton, which vary widely in their properties and characteristics.

The assessment of cotton is inevitably a difficult job, requiring great experience and skill. In general, quality is linked directly with staple length, and this is the characteristic that is commonly quoted in referring to any cotton. Staple length is an assessment of a fibre with respect to its technically most important length. In the case of cotton, staple length corresponds very closely with the most frequent length of the fibres when measured in a straightened condition.

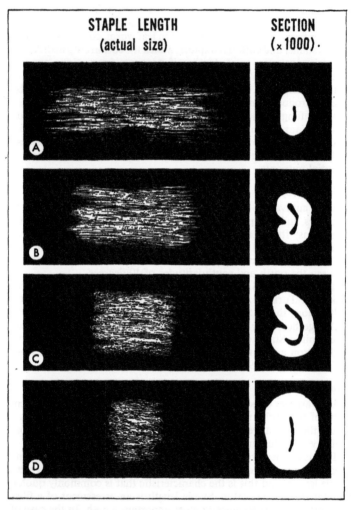

Cotton. Comparison of staple lengths and mature sections of different types of cotton. (A) Sea Island (St. Vincent); (B) Sudan (Gezira); (C) American (Texas); (D) India (Bengals) – *After Cotton Board.*

Commercial cottons may be classified broadly into three categories with reference to the staple length:

(1) STAPLE LENGTH 1–2½ IN. (26–65mm). Includes the fine, lustrous fibres which form the top quality cottons. The fibres are generallly of 10–15 microns diameter 1.1–1.8 dtex (0.99–1.62 den).

Sea Island, Egyptian and American Pima (American-Egyptian) cottons are in this category. These high quality cottons are often the most difficult to grow, and are in comparatively short supply.

(2) STAPLE LENGTH ½–1⁵/₁₆ IN. (12–33mm). Includes the medium strength, medium lustre cottons which form the bulk of the world crop. The fibres are generally of 12–17 microns diameter, and are of 1.4 — 2.2 dtex (1.26–1.98 den).

American upland and some Peruvian types come into this category.

(3) STAPLE LENGTH ³/₈–1 IN. (9–26mm). Includes the coarse, low-grade fibres which are often low in strength and have little or no lustre. The fibres are generally of 13–22 microns diameter, and are of 1.5–2.9 dtex (1.35–2.61 den).

Many of the Asiatic, Indian and some Peruvian cottons come into this category.

Ginning the Seed Cotton

After picking, the cotton fibre has to be separated from the seeds, a process carried out mechanically by the cotton gin. There are two forms of this machine in general use, the saw gin and the roller gin. The saw gin is used mainly for short and medium length cotton, and the roller gin is often preferred for longer fibres, although the short Asiatic types of India and Pakistan are roller ginned. Roller ginning is a slower and more costly process.

The Saw Gin

This consists of a steel grating in which are narrow slits. Through these come toothed saws that revolve, catching the fibres in their teeth and pulling them through the slits. The seeds are too big to go through, and remain behind. The ginned cotton is called 'lint'.

Ginning does not remove all the cotton; short fibres are left adhering to the seeds. These fibres are removed by passing the seed through another gin, and the mass of short fibre produced ('linters') is used for stuffing upholstery and as a source of pure cellulose for industry.

The Roller Gin

This consists of leather discs attached to a wooden roller. The leather surface of the revolving roller passes close to a 'doctor' knife leaving a space through which fibres can pass but seeds cannot. As the roller revolves, the fibres cling to the leather surface and are carried through the gap between the leather and the knife. The seeds are caught by the knife and removed.

Ginned cotton is pressed and packed into bales weighing 200 – 720 lb (91 – 327kg) and sent off for spinning into yarns. 1 óz (28.4g) of cotton contains about 100 million fibres, so that a 500 lb (227kg) bale will contain some 800,000 million fibres. Placed end to end, these fibres would reach 20 million km (12½ million miles).

Grading

Hundreds of varieties of cotton are grown in different climatic conditions and in all manner of soils and environments. The grading and classification of all these cottons, with special reference to the yarns and fabrics they will produce, is obviously no simple task.

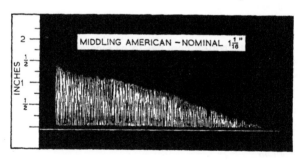

Cotton. This drawing shows a typical staple array of Middling American cotton (nominal 1 1/16 in.) – *After Cotton Board.*

The assessment of cotton is carried out traditionally by the cotton 'classer', who depends upon personal skill and long experience in judging cotton quality by inspection and feel. In arriving at his assessment, the classer takes note of (1) the staple length, (2) the colour and (3) the amount of impurity in the cotton, and the quality of its preparation.

In the U.S.A. and other cotton producing countries, standards have been established, using these factors as the basis of cotton classification, and it is the classer's job to assess a batch of cotton with reference to these standards.

The classer works very largely by a hand-examination of the cotton. Staple length is judged by taking a sample and pulling it to display a filmy web of fibre.

In addition to these basic qualities, a skilled classer will take many other factors into consideration in arriving at his judgment of the value of the cotton to the spinner and the weaver. Nowadays, he can support his personal judgment by laboratory tests which give precise, numerical values to the various properties of the cotton.

Cotton Spinning

Cotton arriving at the mill is normally dirty and contaminated by bits of leaf, dust and twigs. Impurities will commonly amount to about 2–3 per cent of the total weight. The cotton has also been compressed tightly in the bale in order to minimize transport costs.

Before the actual spinning of the cotton can be carried out, the fibres are subjected to several preliminary processes. First, cotton from the bale is put into a machine called an opening hopper, in which several spiked rollers are revolving at high speed. The spikes tear the tightly-packed cotton apart, loosening the fibres and allowing many of the impurities to sift out.

Next, the cotton passes to similar machines in which it is beaten to free it from impurities that remain. Another machine continues the cleaning process, delivering the cotton eventually in the form of 'laps', which are continuous sheets of fibre about 102cm (40 in) wide and 2.5cm (1 in) thick. At this stage, the cotton is like an enormous roll of 'cotton wool'.

Carding

The next process is 'carding', which is carried out by a 'carding engine'. This process separates the cotton fibres, takes out most of the remaining impurities and removes short and immature fibres.

From the carding engine, the cotton is delivered as a filmy web which is collected together to form a loose rope of fibres which are just able to cling together and support their own weight. This loose, soft rope is called a sliver.

51

Drawing

Slivers from the carding engine are stretched or drawn in stages by passing them through a drawing frame. This consists of a series of rollers, like a succession of tiny 'wringers', in which each set of rollers is revolving slightly faster than the previous set. Slivers are fed into the drawing frame in groups of from four to eight; as they pass between the rollers they are drawn out into narrower slivers in which the fibres have been blended and aligned more closely. They are collected together as they leave the drawing frame, forming a single draw frame sliver.

Combing

If the cotton is to be made into fine or high-quality yarns it is combed to align the fibres more accurately and to remove more of the shorter fibres. This operation is carried out in a comber. The sliver from the draw frame is passed through a machine which converts it into small laps; the laps are fed into combers in which the cotton is combed by a revolving cylinder equipped with many needle-like spikes. The cotton comes from the comber as a fine web which is again collected into a loose rope or sliver. Several of the individual slivers are brought together and combined into a single sliver.

The stretching or drawing process is repeated on other machines, such as the slubber and roving frame, until eventually the cotton is in the form of a much narrower sliver in which the fibres are lying sufficiently parallel and uniform to be ready for spinning. At this stage, the sliver acquires a new name; it is a roving if it comes from the roving frame, or slubbing if it comes from the slubber.

As the roving is wound up on leaving the frame it is given a slight twist. This enables the fibres to hold onto one another sufficiently to prevent the roving from breaking.

The number of 'drawing' processes through which the cotton sliver or roving passes depends upon the nature of the yarn that is to be made. The more the cotton is attenuated, the finer and more uniform will the yarn be.

Spinning

All these stages in the processing of cotton have developed with the mechanization of textile manufacture. When spinning was carried out with simple hand equipment, such as the spindle or the spinning wheel, it was not possible to ensure that the yarn would be absolutely uniform. Nor was it necessary when yarn was to be woven on simple hand-operated looms. But mechanization, in textiles as in other industries, demands a degree of uniformity; the processes leading up to the production of the roving are designed to provide a yarn that is consistent and is suitable to act as a raw material for the mechanized weaving processes.

The roving corresponds, in effect, with the strand of fibre that was drawn out from the mass by the hand-spinner. The simple operation of pulling out a strand of fibres in this way provided some degree of alignment to the fibres, and by stretching his strand of fibres during the twisting operation the spinner was able to exercise his skill in order to obtain a reasonably uniform yarn.

Mechanization

The mechanization of this simple spinning process was accomplished only with difficulty, and many inventors, mostly Lancashire men, contributed to the success which was achieved.

In 1738, Lewis Paul of Birmingham invented the method of drawing out cotton slivers until they were fine enough for spinning into yarn. Between 1764 and 1767, James Hargreaves of Standhill developed the spinning jenny on which many threads or yarns could be spun at a time. Then came Richard Arkwright, who perfected the method of using rollers for drawing out the fibres during spinning; Arkwright's inventions enabled the spinner to produce a finer and firmer yarn than had been possible with Hargreaves' jenny.

Between the years 1774 and 1779, Samuel Crompton of Bolton followed up these inventions and combined the principles of Hargreaves' jenny and Arkwright's water-driven frame in a machine called the mule. Cotton could be spun on the mule into a yarn much finer than any that had previously been manufactured.

In mule spinning, which still survives in a few places, rovings are drawn out and mounted as spindles on a moving carriage. As this runs outwards, the spindles revolve, twisting the roving into a yarn or thread which is wound up as the carriage returns in the opposite direction.

The modern spinning process is called ring spinning, which is carried out on a more compact machine without a movable carriage. The operation of the ring spinner is continuous, by contrast with the to-and-fro intermittent movement of the mule. The basic operations of stretching and twisting the roving are, however, the same.

Count

The fineness of the yarn produced by spinning is denoted by a number called the 'count'. This is a measure of the length of a certain weight of the yarn; the finer the yarn, the higher is the count.

The count of a cotton yarn is the number of hanks, each 840 yards (756m) long, in 1 lb (454g). Fine-spun yarns have counts of 50 or more with an upper limit for cotton of about 300. This type of ultra-fine yarn is spun from high-quality Sea Island cotton of the longest and finest staple. A 240 count cotton may be spun from high-quality Sea Island fibre of staple length 1¾ in (44mm).

Most of the world's fine cotton spinning is done in the 50 – 150 count region. Counts of less than 50 are regarded as medium or coarse yarns. A good quality Bengal cotton of only ½ in (12mm) staple, for example, will spin to a count of 12.

To produce a yarn of a desired fineness, the spinner buys cotton of a staple-length and quality that is capable of providing the count he wants. By adjustment of the machines, he can then spin the cotton to this count.

When the spinner has done his job, the cotton fibres have been drawn out and aligned, and then twisted together so that they grip one another and can combine to resist a 'pull'. The finished product provided in this way by the spinner is a yarn.

Yarns may be processed further in many ways. Several yarns are, for example, twisted together or 'doubled' to form threads.

Scouring and Bleaching

Cotton is normally spun into yarns and threads without undergoing any treatment other than the mechanical processes already described. A small amount of mineral oil may be added during spinning, and yarns may be treated later with size. The non-cellulosic materials remain in the cotton, however, and the untreated yarns, threads or fabrics are known as grey goods.

Before these grey goods can be dyed and finished for sale, they are cleaned and purified by processes that are described fairly generally as bleaching. There are many modifications of cotton bleaching techniques in use today; they are carried out largely after the grey yarns have been woven into cloth.

Before bleaching, the cloth is often singed by passing it quickly across open gas flames, or between red-hot rollers or plates. The cloth is then quenched in water or dilute sulphuric acid.

During weaving, the cotton yarns are normally sized with starch. This size is removed after singeing by steeping the cloth in dilute sulphuric acid or with the help of enzymes. The cloth is washed thoroughly after desizing.

Alkali Treatment

Impurities are removed from the cotton cellulose by the process known as kier boiling. The cotton is placed in a pressure vessel and heated in a solution of caustic soda at about 118°C (250°F).

Kier boiling removes some of the wax from the surface of the cotton fibres, and increases the 'wettability' of the cotton. The cotton is washed thoroughly and is then ready for the bleaching proper which destroys remaining coloured non-cellulosic impurities and whitens the cotton.

Bleaching

From the earliest days of the cotton industry, chlorine compounds have been used in bleaching cotton. Bleaching powder or sodium hypochlorite release active oxygen which combines with the impurities without doing appreciable damage to the cotton cellulose.

After bleaching, the cotton is treated with dilute sulphuric acid, washed and then treated with sodium bisulphate to remove residual chlorine. It is then washed thoroughly.

In recent times, hydrogen peroxide has been used increasingly as a cotton bleach in place of the traditional chlorine compounds. Much of the U.S. cotton is now bleached with peroxide; the process saves time and the cloth is subjected to less handling. In addition, the finished cloth has a softer handle.

Chemical Modification of Cotton

The cellulose of cotton is an active chemical, and its character and behaviour can be altered by chemical treatment. Research in recent years has been extremely active in this field, and chemically modified cottons are now appearing on the market in a number of different forms. These cottons retain the essential fibre structure of the original cotton, but the cellulose from which they were originally formed has been subjected to chemical modification. It is no longer cellulose, but a chemical derivative of cellulose.

These changes brought about in the essential chemical structure of the cotton are accompanied by permanent changes in the properties of the cotton. In effect, they provide us with entirely new types of fibre.

Cotton is a particularly useful fibre to serve as the raw material for chemically modified fibres of this sort. It is produced in great abundance, and it can therefore carry the added cost of chemical treatment without becoming too expensive to be of real commercial value.

PA COTTON. Treatment of cotton with acetic anhydride in acetic acid converts it to partially acetylated cotton (PA cotton). This material looks like the original cotton; it has no smell and is non-toxic. But in many of its properties, PA cotton differs from the normal fibre. Most important of all, it has a greater resistance to heat than cotton. At 250°C., for example, cotton loses one-third of its strength in three minutes, whereas PA cotton in similar yarns loses one-third of its strength only after twenty-five minutes at the same temperature. That is to say, PA cotton lasts for eight times as long before losing one-third of its strength. This extra heat-resistance is borne out in practice in the added life of cotton fabrics that are constantly being subjected to heat. Laundry press covers, for example, will last five times as long when made from PA cotton as they do when made from ordinary cotton.

Added to this heat-resistance, PA cotton withstands the attacks of micro-organisms such as those responsible for mildew and rotting. In a test carried out at the Southern Regional Research Laboratory of the U.S. Department of Agriculture, ordinary cotton lost most of its strength after burial in active soil for a week; ordinary cotton treated with copper rot-proofing agents had lost two-thirds of its strength in eight to twelve weeks. PA cotton, on the other hand, retained more than four-fifths of its original strength after being buried for almost a year.

PA cotton is also more resistant to attack by certain chemicals. In 20 per cent hydrochloric acid, for example, it loses only one-third of its strength after eight hours, whereas ordinary cotton loses about two-thirds.

Under certain circumstances, PA cotton shows better weathering resistance than ordinary cotton. It is also a better electrical insulator.

The increasing interest shown in chemically modified cotton has stimulated development of PA cotton. But it is not a 'new' material; PA cottons have been produced in Europe on a small scale since the 1930s.

PA cotton is being used for a number of diverse applications. Sandbags made from PA cotton will last two years under conditions where rot-proofed cotton bags disintegrated in 2–5 months.

PA cotton has been made into fishing nets and lines which were doing useful service after eight months' use; cotton equipment lasted only for a month under similar conditions, and tar-coated cotton was unserviceable after four months.

AM COTTON. When cotton is treated with 2-aminoethylsulphuric acid in sodium hydroxide, another form of chemical modification takes place. Once again, the fibre retains its essential structure, but its properties have changed. The new fibre is known as AM cotton.

AM cotton accepts certain types of dyes more readily than does ordinary cotton; the dyed materials have better resistance to light and washing. The chemical groups in AM cotton are able to react readily with other chemicals, and many new properties can be given to the fibre by subsequent treatment. Rot-resistance can be 'built into' AM cotton in this way.

CM COTTON. Cotton treated with monochloroacetic acid and then sodium hydroxide is converted into CM cotton. Two distinct types can be made in this way.

One type of CM cotton has a 'starched' appearance and handle. It absorbs water more readily than cotton and can accept crease-resisting treatments with greater effect.

The second form of CM cotton disintegrates readily in water. It can be used as a temporary yarn for making fabrics from which the unwanted yarn can easily be removed.

The insoluble CM cotton can be produced easily in mercerizing equipment, at very low cost. The product can be crease-proofed with particularly good effect.

CN COTTON. Treatment of cotton with acrylonitrile yields a chemically modified cotton described as cyanoethylated (CN) cotton. This fibre looks and feels like ordinary cotton, and many of its physical properties are similar to those of cotton. CN cotton, however, has extremely good resistance to rotting influences.

Buried in soil, samples of CN cotton retained their full tensile strength after seventy days; ordinary cotton had rotted completely in five days.

CN cotton has a much better resistance to the effects of heat tnan has ordinary cotton; this resistance is retained under conditions of high humidity.

CN cotton dyes more readily than cotton with certain types of dye.

PL COTTON. Treatment with propiolactone converts cotton into a modified cotton described as PL cotton.

MISCELLANEOUS. The reactive hydroxyl groups of cellulose can be oxidized to aldehyde with periodic acid, hypohalites or nitrogen dioxide; the latter has yielded a fibre used as a soluble surgical dressing.

Cotton treated with phosphoric acid and urea is converted into a phosphorylated cotton containing up to 4 per cent of phosphorus. This fibre, though weaker than normal cotton, is flame resistant. It has cation-exchange properties similar to the ion-exchange resins that are widely used in water-softening.

Ethylene oxide dissolved in carbon tetrachloride will modify the structure of cotton in such a way as to give cotton fabric an organdie-like finish.

The treatment of cotton with formaldehyde confers crease-resistance and affects the dyeing behaviour of the fibre.

Cotton can be made water-repellent by treatment with stearamido-methyl pyridinium chloride. This is the basis of the Zelan and Velan processes used commercially.

Dyeing

Cotton is used for an immense variety of textile applications and dyestuffs are available which will give satisfactory results under almost any conditions encountered in practice.

Vat and azoic dyes are used when the colours must be fast to washing and chlorine. Sulphur colours, developed direct colours and Hydron Blues provide good light and washing fastness. Direct colours are used when the fabrics are not to be subjected to exposure or to repeated washing; they are also used in dyeing union fabrics containing cotton. Basic dyes provide attractive, brilliant shades when fastness is not a primary consideration.

STRUCTURE AND PROPERTIES

Fine Structure and Appearance

One of the characteristics of cotton is the great variety of shape and form that its fibres show. Any sample of raw cotton will contain fibres in different stages of development, and the microscopic features of these fibres will differ widely one from another. Moreover, samples of cotton from different varieties of plant and from plants grown under different conditions will differ too.

Convolutions

The cotton fibre is a single cell which has collapsed into a flattened tube of cellulose as it dried. The mature fibre can be recognized by the twists or convolutions which are a characteristic of cotton. A typical Sea Island cotton will have some 300 half-convolutions to the inch (25mm); Egyptian 230; Brazilian about 210; American about 190, and Indian 150. The direction of twist reverses after every two or three convolutions.

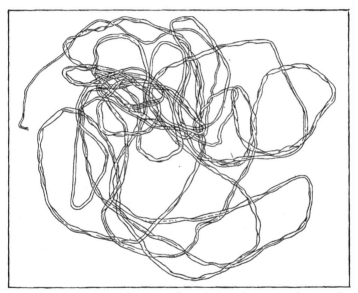

Cotton. A single cotton fibre seen through the microscope, showing the variations in size and shape of the dried, shrivelled tube – *After Mary L. Rollins, U.S. Dept. of Agriculture,*

Length

The length of the individual cotton fibre varies greatly, depending upon the variety of the plant, the conditions under which it has been grown, and the state of maturity of the fibre at the time of picking. A good Sea Island fibre may be 2½ in (65mm) long, whereas a linters fibre will be less than ¼ in (6mm) in length.

Fineness

In general, the cotton fibre is of fairly uniform width. At one end, it tapers to a tip. The other end of the fibre is open and irregular; this is the point at which the fibre was torn from the seed during ginning.

The actual width of a typical cotton fibre 'ribbon' may vary between about twelve and twenty microns.* The central section of the fibre is thicker than the ends.

Seen in cross-section, the normal mature fibre is oval or kidney-bean shaped. The thick cellulose wall encloses a well-defined lumen.

Immature Fibres

Immature cotton fibres can be recognized by the thinner cell walls. The dry, immature fibre does not show the oval or kidney-bean shape cross-section; instead, the fibre tube collapses into a thin ribbon that curls into a variety of distorted shapes. Often the immature fibre is U-shaped in cross-section.

These thin-walled, immature fibres may not twist as they collapse on drying. They seldom have the pronounced convolutions that are so typical a feature of the mature fibre.

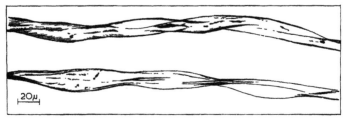

Cotton. Longitudinal view of mature (above) and immature (below) cotton fibres, showing typical convolutions – *After U.S. Dept. of Agriculture*

Micro-structure

In recent years, studies have been made of the fine structure of the cotton fibre. New techniques and instruments such as the electron microscope have enabled us to examine the internal make-up of the fibre in a way that has not been possible before.

The wall of the fibre varies in thickness. It consists of two main sections, the primary wall or cuticle forming the outer layer, and the secondary wall forming the inner layer.

* 1 micron = 1/1,000 mm. = 1/25,000 in. approx.

Primary Wall

The primary wall is a tough, protective layer that formed the shell of the fibre during its early days of growth inside the boll. Seen under the microscope at high magnification, the surface of the cotton fibre is wrinkled like a prune. These surface wrinkles are caused by shrinkage as the fibre has dried.

At the Southern Regional Research Laboratory of the U.S. Department of Agriculture, methods have been devised for removing the primary wall from the fibre. Cotton is beaten in water under special conditions and short fragments of the primary wall slip off like sleeves. Chemical analysis of primary wall material isolated in this way has shown that it contains wax, protein and pectinaceous substances as well as cellulose. When these non-cellulosic materials are removed chemically, the cellulose fibrils can be seen with the help of the electron microscope as a felt-like mat of tiny threads. These cellulose fibrils are between 1/40th and 1/100th of a micron in diameter, and consist of many long cellulose molecules held tightly alongside one another by natural forces of attraction that are exerted between close-packed molecules.

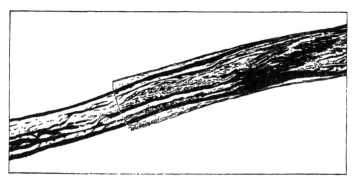

Cotton. In this photomicrograph, the outer membrane of the cotton fibre is seen peeling from the fibre surface – *After Mary L. Rollins, U.S. Dept. of Agriculture.*

These tiny fibrils of cellulose are, like the fibre itself, strongest in a longitudinal direction. The criss-cross network of fibrils forming the primary wall of the cotton fibre is an arrangement that confers great peripheral strength. The outer wall of the fibre is able to resist forces from any lateral direction, although it does not have the immense longitudinal strength that would be available if all the cellulose fibrils were arranged side by side along the longitudinal axis of the fibre.

This tough-skin effect of the primary wall is noticeable when cotton is immersed in solutions which are able to swell and dissolve cellulose. In cuprammonium hydroxide, for example, the cotton fibre swells up into a distended tube which is tied at intervals like a string of sausages. This is probably due to the resistance of the primary wall to the effects of the solvent. As the cuprammonium hydroxide penetrates and swells the secondary cellulose inside the fibre, the primary wall ruptures and is rolled back into tight ligatures that resist the swelling effects of the solvent. This resistance on the part of the primary wall is largely due to the matted arrangement of the cellulose fibrils forming the network of the wall. In this respect, it is interesting to find that immature fibres, which have little secondary cellulose, are swelled only slightly by solvents.

Secondary Wall

The inner or secondary layer of cellulose forms the bulk of the cotton fibre. This is the cellulose that is laid down during the second stage of fibre growth, after the fibre has attained its full length, when consolidation of the cellulose wall takes place.

The growth-ring arrangement of the cellulose in the secondary wall can be seen when swollen fibres are examined microscopically. With the help of the electron microscope, the organization of the fibrils can be followed. The fibrils of the secondary wall are packed together in a near-parallel arrangement. The layers of fibrils lie in spiral formation along the fibre, the direction of the spirals often reversing in the same layer.

The secondary wall is almost pure cellulose, and represents about 90 per cent of the total fibre weight. The bulk of the cellulose in the cotton fibre is therefore arranged in the form of fibrils that are packed alongside each other; they are aligned as spirals running lengthwise through the fibre. The fibrils are therefore able to exert high strength in a longitudinal direction in the fibre bestowing immense longitudinal strength on the fibre itself.

Lumen

When the cotton fibre is alive and growing, it is distended by the pressure of the liquid nutrients and protoplasm inside it. As the fibre dies and collapses, the liquid disappears leaving an almost

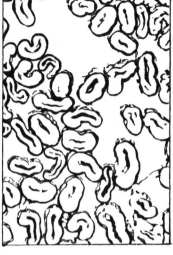

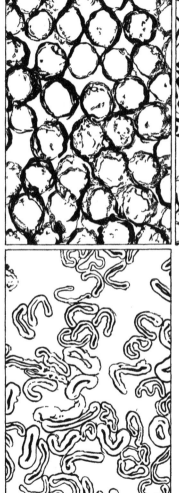

Cotton. Cross sections of cotton fibres.

Top Left: Growing fibres filled with plant juices are round in cross section.

Top Right: Thick-walled, mature fibres collapse into oval shapes on drying.

Bottom Left: Thin-walled, immature fibres collapse into distorted shapes on drying – *After Mary L. Rollins, U.S. Dept. of Agriculture.*

empty channel running lengthwise through the centre of the fibre. This central canal is the lumen.

In a mature cotton fibre, the deposition of cellulose in the secondary wall may have been so heavy as to leave very little lumen at all. The dry fibre looks like a solid rod rather than a tube. An immature fibre, on the other hand, may have so little secondary cellulose that the lumen is wide and distinct.

When the cellulose of the primary and secondary walls is dissolved by powerful solvents, a thin membrane of protoplasm is left behind. This is the dried-up residue of the materials that were dissolved in the watery liquid inside the living fibre. It contains a coloured substance, the endochrome, which gives the cotton its natural colour. Cottons have been grown experimentally in which the lumen contains brown or green pigments which act as natural 'dyes'.

In a normal cotton fibre, the lumen although almost completely collapsed represents a considerable volume of unoccupied space. It enables the cotton fibre to absorb water by capillary attraction and so has an important influence on the properties of cotton as a textile.

The lumen, however, forms only a part of the unoccupied space in a cotton fibre. The cellulose walls are porous and can absorb considerable quantities of water on their own account.

Although the fibrils of cellulose forming the fibre walls are compact and relatively impervious to water penetration, the submicroscopic spaces between them form capillaries that make the cellulose network porous. Large surfaces are exposed by the fibril network, and water is able to penetrate by capillary attraction.

It has been estimated that as much as 20–41 per cent of the volume of a cotton fibre consists of unoccupied space. One-third of this is accounted for by the lumen; the rest is provided by the spaces between the fibrils in the fibre walls. In general, the coarser varieties of cotton are more porous than the compact, finer cottons.

Fibre Surface and Colour

The surface of the cotton fibre, seen at high magnification, is wrinkled and striated. But for most practical purposes, cotton can be regarded as having a smooth surface.

When cotton is spun to form a yarn, the fibres are able to hold together, despite this smoothness, by virtue of their convolutions. The natural twists enable the fibres to grip one another and prevent the slippage that would otherwise take place. A maximum grip is exerted by fibres with 150–175 half-convolutions to the inch (25.4mm).

Cotton fibres have a natural lustre which is due, in part, to the natural polish on the surface. The smooth, hard primary coat of cellulose contains waxes which no doubt contribute to the lustre of the fibre. This surface-smoothness, however, is not the only factor that controls the lustre of cotton. The shape of the fibre is important as well; a high lustre is provided by fibres of nearly circular cross-section and with fewer convolutions such as those produced when cotton is mercerized.

The colour of cotton, normally creamy-white, is affected greatly by the conditions under which it is produced. If the fibre is left too long in the boll before being picked, it may turn grey or bluish-white. A sharp frost will sometimes open the bolls prematurely: this cotton is often darkened to a buff colour. Its fibres are immature and weak.

Tensile Strength

Cotton is a moderately strong fibre; tenacity is 26.5–44.1 cN/tex (3.0–5.0g/den) and tensile strength 2800–8400kg/cm^2 (40,000–120,000 lb/in^2). The strength is affected greatly by moisture (see below) and by the test conditions such as rate of loading, and length of fibre section tested.

The long, fine cottons, such as Sea Island and Egyptian, yield the strongest yarns and materials. Yarn strength is a complex property that involves many factors. The long cotton fibres, for example, are able to grip one another more effectively than the shorter ones, so that there is less tendency for slippage to take place. Long fine fibres of good strength can thus be spun into finer yarns.

Elongation

Cotton does not stretch easily. It has an elongation at break of 5–10 per cent.

Elastic Properties

Cotton is a relatively inelastic, rigid fibre. At 2 per cent extension it has an elastic recovery of 74 per cent; at 5 per cent extension, the elastic recovery is 45 per cent.

Specific Gravity. 1·54.

Effects of Moisture

The tensile properties of cotton fibres and yarns are affected appreciably by the amount of moisture absorbed by the fibres.

Under average humidity conditions, cotton takes up about 6–8 per cent of moisture; it has a regain of 8·5 per cent. At 100 per cent humidity, cotton has an absorbency of 25–27 per cent.

Up to a relative humidity of 100 per cent, absorption of water by the cotton cellulose results in an increase in fibre strength. Fibres saturated with water are about 20 per cent stronger than dry fibres. Cotton yarns will continue to become stronger at high relative humidities.

The humid atmosphere of Lancashire was peculiarly favourable to the spinning and weaving of cotton during the early years of the industry. The damp yarns and threads were not so inclined to break during manufacture. Today, humidity in a factory can be controlled artificially as easily as temperature, and the climate of Lancashire is no longer a critical factor in the cotton trade.

Action of Water on Cotton

The dry cotton fibre, constructed from its fibrils of cellulose, is a fairly stiff, rigid entity. The cellulose molecules are held tightly together inside the fibrils, bound by bonds established between molecules lying close alongside one another.

Water, however, is able to penetrate into the cellulose network of the cotton fibre. It makes its way into the capillaries and spaces between the fibrils and into less tightly bound areas of the fibrils themselves and attaches itself also by chemical links to groups in the cellulose molecules. In this way, water molecules tend to force the molecules of cellulose apart, lessening the forces that hold the cellulose molecules together and destroying some of the rigidity of the entire cellulose structure.

Water acts, in this way, as a 'plasticizer' for cotton. By penetrating into the mass of tightly-bound cellulose molecules it permits the molecules to move more freely relative to one another. The mass of cellulose is softened, and can change its shape more easily under the effects of an applied force.

This action of water on the cotton fibre is responsible for a number of important features in the processing of cotton fabrics. The cellulose molecules in wet cotton are so well lubricated by the water molecules that the fibres become quite plastic and easily deformed. And the effect of pressure upon the cotton is identical with its effect upon any other plastic – it changes its shape. When cotton fabrics in this state are passed through a pair of pressure

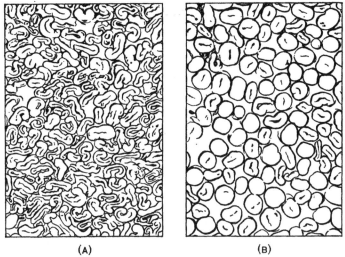

(A)   (B)

Mercerization. During mercerization, the cellulose of the cotton fibre is swollen rapidly, and the fibre becomes almost a solid cylinder of cellulose. The effect remains even after the cotton has been washed and neutralized. The drawing shows cross sections of cotton yarns; (A) untreated; (B) mercerized – *U.S. Dept. of Agriculture.*

rolls, therefore, the fibres accommodate themselves to the force applied to them; a smooth, flat finish is given to the cloth.

The ability of cotton fibres to take up water has been used in a number of other ways. Tightly woven cloths, for example, can be made in such a way that as soon as the fibres become wet they absorb water and swell. As they swell, the cotton threads close up the interstices of the cloth, thus preventing water from getting through.

The swelling of cotton yarns and fabrics in water is accompanied by some shrinkage.

Mercerization

A most valuable application of this swelling of cotton is in the process known as mercerization. Here, a strong solution of caustic soda is used to bring about intensive swelling of the fibres so that the cotton can be deformed.

Mercerization was discovered in 1844 by one of Britain's great textile scientists – John Mercer of Accrington – who found that by treating cotton with caustic soda the fibres could be made to swell. This caused an overall shrinkage in the fabric as the strains between the threads released themselves. At the same time, the fabric became much stronger and was much more easily dyed.

The mercerization process was taken a stage further in 1890, when H. A. Lowe found that by holding the cotton so as to prevent its shrinking during mercerization it developed the beautiful lustre that we now associate with the process.

During mercerization under these conditions the cotton fibres regain their original circular cross-section due to the swelling of the cellulose, and they tend to lose their convolutions. As they are held fast the whole time, they cannot take up the strains by shrinking; the plastic cellulose deforms and the cloth develops a smoother surface than it had before.

After washing and drying, the mercerized cotton fibre retains this smooth cylindrical form.

Nowadays, cotton yarn is mercerized usually to improve its lustrous appearance and its dyeing qualities. Cloth is mercerized to obtain similar effects and to stabilize its dimensions. When the cotton contains a high proportion of thin-walled immature fibres, mercerizing will swell these fibres and make them dye more like maturer fibres.

Mercerized cotton is chemically little different from ordinary cotton. It remains almost pure cellulose, but in a different physical form. It is more reactive, and will take up about 12 per cent of water from the air, compared with its usual 6–8 per cent.

Sulphuric and other acids are being used to swell cotton and yield effects that resemble those of mercerization. Cotton is passed rapidly through strong sulphuric acid, so that the material is in contact with the acid for only a few seconds. Acid finishing of this sort is adapted to provide a number of different effects. Cotton fabric is made more transparent or more wool-like in this way. It is given an organdie finish which is permanent; the fabric regains its characteristic stiffness after washing and ironing.

Many other acids and salts are used for treating cotton, and the details of such processes are often undisclosed. The effect produced depend on the conditions used and on the nature of the fabric itself. In most cases, however, it is the swelling of the cotton fibre that is the most important factor.

Effect of Heat

Cotton has an excellent resistance to degradation by heat. It begins to turn yellow after several hours at 120°C., and decomposes markedly at 150°C., as a result of oxidation. Cotton is severely damaged after a few minutes at 240°C.

Cotton burns readily in air.

Effect of Age

Cotton shows only a small loss of strength when stored carefully. It can be kept in the warehouse for long periods without showing any significant deterioration. After fifty years of storage, cotton may differ only slightly from fibre a year or two old. Ancient samples of cotton fabric taken from tombs more than 500 years old had four-fifths of the strength of new material.

Effect of Sunlight

There is a gradual loss of strength when cotton is exposed to sunlight, and the fibre turns yellow.

The degradation of cotton by oxidation when heated is promoted and encouraged by sunlight. It is particularly severe at high temperatures and in the presence of moisture. Much of the damage is caused by ultra-violet light and by the shorter waves of visible light. Under certain conditions, the effects of weathering in direct sunlight can be serious. The cotton can be protected to some degree by using suitable dyes.

Chemical Properties

Cotton fibre, as it is picked from the plant, is about 94 per cent cellulose. The remaining 6 per cent is made up of protein (1–1·5 per cent), pectic materials (1 per cent), mineral substances (about 1 per cent), wax (about 0·5 per cent) and small amounts of organic acids, sugars and pigments.

Much of this non-cellulosic material is removed from cotton by scouring and bleaching processes, leaving a fibre that consists of about 99 per cent cellulose.

The purification of cotton cellulose provides a stronger and whiter fibre which absorbs moisture more readily. The natural wax has a useful lubricating effect during spinning, however, and most cotton is spun with its wax still present in the fibre.

Analyses of fibres from different sources have shown that the amount of wax, pectin and protein increases with increasing immaturity of the fibre.

Cotton is highly resistant to the chemicals encountered in normal use. Dyestuffs, mild bleaching agents and similar materials have no significant deleterious effects on cotton fabrics if used with reasonable care. Cotton is attacked by strong oxidizing agents, including hydrogen peroxide and chlorine bleaching compounds.

The purity of scoured and cleaned cotton, and the chemical stability of cellulose, together make cotton into a remarkably durable material.

Effect of Acids

Cotton is attacked by hot dilute acids or cold concentrated acids, in which it disintegrates. It is not affected by cold weak acids.

Effect of Alkalis

Cotton has an excellent resistance to alkalis. It swells in caustic alkali (cf. mercerization) but is not damaged. It can be washed repeatedly in soap solutions without taking harm.

Effect of Organic Solvents

There are very few solvents that will dissolve cotton completely. It has a high resistance to normal solvents but is dispersed by the copper complexes cuprammonium hydroxide and cupriethylene diamine, and by concentrated (70 per cent) H_2SO_4.

Insects

Cotton is not attacked by moth grubs or beetles.

Micro-organisms

Cotton is attacked by fungi and bacteria. Mildews, for example, will feed on cotton fabric, rotting and weakening the material. They have a characteristic musty smell, and stain the fabric with naturally produced pigments.

Mildews are particularly troublesome on cotton that has been treated with starchy finishes, and much of the damage can be avoided by thorough scouring. The pure cellulose is a less attractive food for mildew than the starch.

Mildews and bacteria will flourish on cotton under hot, moist conditions. When cotton fabrics are to be used under conditions favourable to attack by micro-organisms, they can be protected by impregnation with certain types of chemical. Copper compounds, such as copper naphthenate, will destroy organisms that would otherwise attack the cotton cellulose.

COTTON IN USE

In cotton, nature has given us an all-round utility fibre that is second to none. Cotton fabrics combine remarkable durability with attractive wearing qualities. Cotton fabrics have a pleasant feel or 'handle'. They are cool in hot weather.

Cotton is inherently strong, and it is stronger when wet than it is when dry. This property, allied with cotton's stability in water and alkaline solutions, endows cotton garments with a long useful life. Cotton can withstand repeated washings, and is therefore ideal for household goods and garments that can be laundered time and time again. Heavily soiled garments can be rubbed vigorously without being damaged.

The cotton fibre itself is dimensionally stable. A made-up cotton garment may shrink to some extent due to the tensions introduced by spinning and weaving, but the fibre itself does not contribute significantly to any shrinkage.

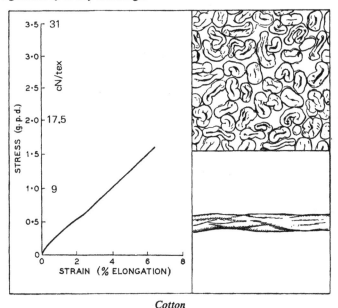

Cotton

71

This so-called 'relaxation shrinkage', caused by the easing of strains set up during spinning and weaving, can be overcome by a treatment called compression shrinkage. 'Rigmel'- and 'Sanforized'-shrunk cotton fabrics are compression-shrunk in this way; they are dimensionally stable and will neither stretch nor shrink more than 1 per cent in either direction.

The resistance of cotton to washing and wear is matched by the permanence of many cotton dyes. Cotton can be dyed easily, and the colours will often remain fast to repeated washings and to prolonged wear. Vat dyes, in particular, are used for cotton goods where first-rate fastness is essential.

Cotton cellulose is not affected unduly by moderate heat, so that cotton fabrics can be ironed with a hot iron without damage. Cotton fabrics come up crisp and fresh on ironing.

The strength of cotton fibre is one of the main factors in cotton's hard-wearing qualities. But fibre-strength alone does not confer hard wear on a fabric. Other factors come into play, such as the ability of fibres to grip one another during spinning. The twists and convolutions in cotton fibres enable cotton yarns to resist being pulled apart.

The cotton fibre is fairly rigid and stiff, and cotton yarns and fabrics are not as flexible, in general, as fabrics woven from wool, nylon or some rayons. They are, however, more flexible than linen.

To achieve flexibility, cotton can be spun into fine yarns and made into a tight-woven fabric. Poplins, voiles and flannelettes are made in this way; they are extremely flexible and yet have the crispness associated with cotton. Heavier fabrics, such as drills and denims, are much less flexible.

Although cotton is used in great quantity as a fabric for hot-weather wear, it is able to provide warmth as well. The warmth of a garment depends very largely on the pockets of air that are entrapped between the fibres in the fabric. Woollen garments excel in this respect, as the wool fibres are crinkly and rough-surfaced; they can be spun and woven or knitted into full fabrics that hold innumerable air-cells which act as insulators.

Cotton fibres are smoother, stiffer and straighter than wool and they do not make up so readily into air-entrapping fabrics. But special cellular weaves can be used to create the air-cells that provide warmth, and the surface of cotton fabrics can be raised to form an air-filled pile. Molletons, winceyettes and flannelettes are cotton fabrics of this sort.

Much of the comfort of a textile material depends upon its ability to absorb and desorb moisture. A garment that does not absorb any moisture at all will tend to feel clammy as perspiration condenses on it from the skin. Cotton fibres, however, are able to absorb appreciable amounts of moisture, and having done so they will get rid of it just as readily to the air. Cotton garments are therefore comfortable and cool, passing on the perspiration from the body into the surrounding air. No matter how tightly woven a cotton fabric may be, it will permit the body to breathe in this way.

This absorbency of cotton makes it an excellent material for household fabrics such as sheets and towels too.

Cotton is widely used in making rainwear fabrics. It can be woven tightly to keep out the driving wind and rain, yet the fabric will allow perspiration to escape. Special rainwear materials are woven in such a way that water swells the cotton fibres and closes up the interstices in the cloth. Ventile fabrics, for example, are close-woven cotton materials of this sort which are given additional water resistance by a chemical proofing with Velan.

The versatility of cotton has made it into the most widely used of all textile fibres. Cotton is made into every type of garment and household fabric. It goes into boots and shoes, carpets and curtains, clothing and hats. Heavy cotton yarns and materials are used for tyre cords and marquees, tarpaulins and industrial fabrics of all descriptions.

MISCELLANEOUS SEED AND FRUIT FIBRES

1. **Coir** (*Cocos nucifera*)
   SOURCE: Coir is a coarse fibre which comes from the husks of coconuts.
   CHARACTERISTICS: Coarse. Dark brown colour. Individual fibres short; 0.5mm (1/50th inch) long. Thick walled with irregular lumen. Surface covered with pores.
   USES: Cordage, matting, brushes.

2. **Tree Cotton** (Malvaceae, *Bombax*). (Vegetable Down or Bombax Cotton).
   SOURCE: Seed fibre of cotton-tree plant belonging to *Bombaceae* family. Grows in tropical countries. Cultivated in West Indies and Brazil.
   CHARACTERISTICS: Soft near-white fibre. Weaker than ordinary cotton. Poor resilience. Individual fibres 25mm (1 in). Cell walls thin and irregular. Circular cross-section. Contains some lignin.
   USES: Wadding and upholstery material.

73

3. Java Kapok (*Ceiba pentandra*)

SOURCE: Seed fibres similar to bombax cotton. Grows in Malaya, Indonesia. Fibre removed by hand from bolls. Dried and shaken up to remove seeds.

CHARACTERISTICS: Extremely buoyant. Soft but inflexible fibre. Too brittle to spin. Fawn colour. Good lustre. Very light weight. Individual fibres up to 32mm (1¼ in). Smooth surface. Transparent. Oval or circular cross-section. Thin walled and wide lumen. Contains lignin.

USES: Padding and stuffing of upholstery. Lifebelts and other safety equipment at sea. (Kapok will support more than thirty times its own weight and does not waterlog.)

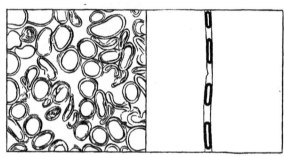

Kapok

4. Balsa Fibre

SOURCE: *Ochroma pyramidale*. Grown in West Indies.

CHARACTERISTICS: Dark brown fibre. Individual fibres up to 12mm (½ in). Folded. Striated surface. Cell wall thick, particularly towards ends. Lumen contains granules. More lignin than bombax cotton.

USES: Stuffing of mattresses and cushions.

5. Kumbi (galgal)

SOURCE: *Cochlospermum gossypium*. Grown in India.

USES: Upholstery and cushion stuffing.

6. Chorisia speciosa

SOURCE: Brazil.

CHARACTERISTICS: Fine silky fibre.

USES: Mattress and pillow stuffing. Spun and woven into fine quality fabrics.

7. Beaumontia grandiflora

CHARACTERISTICS: Pure white fibres with good lustre. Very strong. Individual fibres up to 50mm (2 in) long. Thin cell walls contain fine pores.

USES: Stuffing upholstery.

8. Strophanthus spp.

Similar to *B. grandiflora*, but yellow colour.

9. Milkweeds (*Asclepias*)

SOURCE: Plants of genus *Asclepias*, *A. syriaca* (common milkweed) and *A. incarnata* (butterfly-weed) give excellent seed floss. Indigenous to America. Perennials which grow well in many types of soil in U.S. Pods harvested in late autumn, ginned to remove seeds (see cotton).

CHARACTERISTICS: 'Vegetable silk' or 'Milkweed floss' has a high lustre and is soft and pleasant to the touch. Yellowish white colour. good buoyancy. Too brittle to spin easily. Individual fibre has thick ridges running lengthwise, sometimes five in each fibre, which distinguish it from bombax cotton. Fibres are single cells, often 26mm (1 in) in length. Small amount of lignin and much oily material.

USES: Substitute for kapok in lifebuoys, etc. Upholstery padding.

10. Calotropis Floss ('Akund')

SOURCE: *Calotropis gigantea* and *C. procera*. Indigenous to Southern Asia and Africa. Cultivated in South America and West Indies. Seed floss harvested by hand.

CHARACTERISTICS: Yellowish floss. Individual fibres are thin-walled cells. Up to 32mm (1¼ in) long.

USES: Upholstery stuffing.

11. Cattail Fibre (Typhaceae)

SOURCE: *Typha latifolia* and *T. angustifolia* (Cattails). Indigenous to many regions of America. Fibres separated mechanically from gathered spikes.

CHARACTERISTICS: Delicate, soft fibre. Greyish colour. Good sound and heat insulation. Buoyant and very light. Individual fibres up to 26mm (1 in). Attached to one another like parachute strings. Fibres round in cross-section. High lignin content makes them brittle.

USES: Substitute for kapok in life jackets, etc. Heat and sound insulation.

TECHNICAL NOTE

Cellulose, the basis of all plant fibres, is a substance of empirical formula $(C_6H_{10}O_5)_n$. It is a polymeric material formed by condensation of glucose molecules in the following way:—

The molecular weight of cellulose appears to vary widely, depending on its source. Cellulose from cotton has been quoted as M.W. 200,000 to 400,000, and cellulose from ramie at 240,000 to 320,000. There are probably between 2,000 and 3,000 glucose residues in the molecule. Some estimates put the figure even higher; cotton cellulose may contain as many as 10,000 glucose residues per molecule.

Micelles; Fibrils

The long, thread-like cellulose molecule is the basic unit from which natural plant fibres are constructed. The association of cellulose molecules takes place in stepwise groupings, fibre-like bodies of increasing size building up into the comparatively massive structure of the fibre itself.

The cellulose molecule is able to align itself alongside neighbouring molecules in such a way that molecular bundles are held together by natural cohesive forces. This alignment or orientation of the cellulose molecules in a regularized fashion confers crystalline properties on the tiny groups of molecules, or micelles.

In a plant fibre, the micelles align themselves together to form micro-fibrils. These sub-microscopic filaments, formed from chains of micelles, are in turn constructional units which align and orientate themselves to form fibrils. The electron microscope enables us to 'see' how these fibrils of cellulose are built up to form the fibre itself. In cotton, for example, the fibrils are laid down in a spiral fashion inside the massive secondary wall of the fibre.

Crystalline-Amorphous Structure

According to modern theories of cellulose fibre structure, the long cellulose molecules can individually form part of two or more crystalline regions. These ordered regions are thus held together by cellulose molecules which run from one crystalline region to another. Between the crystalline regions the cellulose molecules are arranged in random fashion, forming regions of amorphous cellulose.

The relative amounts of crystalline and amorphous cellulose have an important influence on the properties of the cellulose fibre. Water is able to penetrate easily between the molecules in the amorphous region, for example, whereas it finds difficulty in entering the close-packed crystalline micelles. Dyestuffs and other substances will tend to favour the regions of amorphous cellulose, leaving the interior of the micelle untouched. Reactions such as acetylation take place more rapidly in cotton than in the more highly orientated flax or ramie. The reagents cannot penetrate so easily into the latter.

Estimates of the proportion of crystalline cellulose in different fibres are difficult to make with accuracy. It is believed that ramie fibre has only a small proportion of random amorphous cellulose, possibly about 5 per cent. Cotton has 5–15 per cent. Other estimates put the proportions of amorphous cellulose very much higher.

Orientation Pattern

The arrangement of the micelles, microfibrils and fibrils with respect to the fibre itself are complicated. In flax and ramie, the micelles are probably orientated almost parallel to the long fibre axis. In cotton, they take up their spiral form.

The pattern of orientation of the molecular bundles has an important bearing on the strength and extensibility of the fibre. The more perfect the orientation, the less extensibility the fibre will show. The micelles are able to hold tightly together, and there will be little 'give' before the fibre ruptures. The mutual grip of the micelles will, however, confer great tensile strength.

If the degree of orientation of the micelles or fibrils is low, a pull on the fibre will tend to align the micelles more accurately. The movement of the micelles as they swing into line confers a high degree of extensibility on the fibre. The tensile strength, on the other hand will be lower than that of a fibre in which orientation is high.

Chemical Reactions of Cellulose

Cellulose is an active chemical, with three hydroxyl groups attached to each glucose residue. Those in the 2 and 3 positions behave as secondary alcohols; the hydroxyl in the 6 position acts as a primary alcohol.

These hydroxyl groups take part in normal chemical reactions, and a great number of cellulose esters and ethers have been made. Some, like cellulose nitrate, are of immense commercial importance.

The chain of atoms forming the cellulose molecule can be broken by chemical attack. Acid hydrolysis produces hydro-celluloses.

Oxidation of cellulose gives rise to oxycelluloses. These are of two types.

(a)   (b)

(a) The acidic type of oxycellulose is formed by alkaline oxidizing agents which attack the carbon atoms in the 6 position on the glucose residues. The alcohol group is oxidized first to aldehyde and then to acid.

(b) The reducing type of oxycellulose, in which cleavage of the ring takes place between the carbon atoms in the 2 and 3 positions.

B: NATURAL FIBRES OF
ANIMAL ORIGIN

Introduction

Animal fibres make up less than 7 per cent of the total weight of textile fibres produced annually. In quantity, therefore, they represent a minor part of the world's fibre resources. But animal fibres play a much more important role in the textile trade than their limited production indicates. They are all fibres of character; each one has unique properties which ensure it a position of special significance as a textile fibre.

Wool and Hair Fibres

Whenever fur-bearing animals form part of the domestic economy of a country, the fibrous materials of their coat are put to good use in one way or another. Wool, the fibrous covering of the sheep, is by far the most important of these fibres, and sheep-farming is now an extremely important activity in many parts of the world. Wool forms more than 90 per cent of the total world production of animal fibres.

Although wool plays such a dominant role in the animal fibre industry, a number of other animal fibres are of considerable commercial importance. In the textile industry, it is usual to describe all animal-covering fibres other than wool as hair fibres. The term wool is restricted to the covering of the sheep. This terminology can lead to some confusion, as the two terms 'hair' and 'wool' are often used to differentiate between the two types of fibre commonly forming the covering of animals. The long, coarse fibres forming the outer coat are called hair, and the short, fine fibres of the undercoat are called wool. It is preferable, therefore, to qualify the term wool by the name of the animal when it refers to anything other than sheep's wool.

Hair fibres are all related to wool in their chemical structure; they are all keratin. But they all differ from wool, and from each other, in their physical characteristics; they are of different length and fineness, and have different shapes and internal structures.

Many hair fibres are used in high-quality applications in the textile trade. Others have specialized non-textile uses; horse-hair, for example, is a padding or filling material; camel hair and pig's bristles are made into brushes; rabbit fur is used for producing felts.

Silk

This fibre, spun by the silkworm as it makes its cocoon, is the only natural fibre of importance which is in the form of a continuous filament. The quantity of silk produced is so small as to amount to only about 0·25 per cent of the total fibre production. But silk has always held a special position as a quality fibre, and sustains a high price by comparison with other fibres.

WOOL

Though vegetable fibres were probably the first to be used for spinning and weaving into cloth, animal fibres in the form of skins and furs were undoubtedly the earliest form of clothing used by primitive man. But at what stage did he discover that the hairs on a sheepskin could be cut off, twisted into yarn and then woven into cloth?

Nobody knows. What we do know is that by the seventh century B.C. the Phoenicians were buying woollen homespuns from the Israelites and shipping them to England to barter for tin and raw wool.

Britain, therefore, first comes into the story of wool as producer of the raw material which was exported to other lands for manufacture into cloth. The small, wild sheep with black faces and long horns that supplied the wool were probably indigenous to Britain. They carried an undercoat of short, fine wool together with a long hairy outer coat.

Domesticated sheep came to Britain with the Celtic tribes who invaded the country during the sixth and seventh centuries B.C. By the time Caesar arrived in 55 B.C. a woollen industry was already flourishing in Britain. Wool was being produced, spun into yarn and woven into cloth.

The Romans built a weaving factory at Winchester, from which they sent home 'wool so fine it was comparable to a spider's web'. The industry developed under Roman rule, but the coming of the Saxons in the fourth and fifth centuries A.D. put an end to the woollen trade by scattering the flocks and destroying the factories.

Then in 1066, William the Conqueror brought fresh supplies of sheep with him from the Continent, and Britain's wool industry re-established itself. The first guild of weavers was established in 1080. This time the industry became centred on Bristol and Exeter; and in 1111, Henry I built up a woollen industry at the mouth of the Tweed. The characteristic heavy woollen cloths that were produced still carry the name today.

By the thirteenth century, another famous name had become familiar in the wool industry. This was the town of Worstead near Norwich – whence came the cloth we now know as 'worsted'. But by this time, the centre of the wool trade had moved to Flanders; Flemish craftsmen were acknowledged experts in the arts of wool weaving and finishing. It was said that 'all Europe is clothed with English wool bought in Flanders'. Raw wool was being shipped to Flanders, made up into cloth and reshipped to England for retail sale.

In 1331, Edward III – at the suggestion of his wife Philippa, a Netherlands woman – invited the Flemish weavers to Britain. Many of these expert craftsmen came and settled in the Norwich area. From that time on, the British wool trade developed and eventually became the leader in the industry – a position it still holds today.

The Woolsack

So important did wool become as a part of Britain's economy that in 1350, Edward III decreed that the Lord Chancellor must sit on a woolsack so that he would remember the importance of the industry during his deliberations. The woolsack is still there.

Further encouragement to the wool-growing industry came from the Black Death that swept over Europe during the fourteenth century. Faced with a severe shortage of labourers – one-third of Britain's population had died – the farmers could no longer plough and cultivate the soil, and they turned increasingly to sheep-raising as being more economical of labour. Britain's sheep population increased until production of wool became our most important industry. In 1454, Parliament declared that 'the making of cloth within all parts of this realm is the greatest occupation and living of the poor commons of this land'.

During Elizabeth's reign, 80 per cent of Britain's exports were wool goods. Every effort was made to encourage the use of wool in the home market. The knitted stocking began to replace the old cloth stockings sewn up from woven material. It was decreed that everyone over seven years of age must wear a wool cap when out of doors; the fine for offenders was 3s 4d – about half a week's wages for a skilled workman.

In his efforts to encourage the use of wool, Charles II carried on where Elizabeth left off. Not content with making his subjects wear their wool caps, he insisted that all women must wear flannel next to the skin. Another decree proclaimed that all corpses should be buried in wool shrouds. This law remained in force until the eighteenth century. And the export of sheep from Britain remained an offence punishable by death until just over a hundred years ago.

This persistent and determined effort to establish the British wool industry was accompanied by a corresponding interest in sheep breeding. Robert Bakewell, born in 1725, became the master-breeder of Britain's sheep; he evolved methods of selection that have given us most of our present breeds. By the end of the eighteenth century Bakewell's Leicestershires were known the world over.

Meanwhile, the inventions of Hargreaves, Crompton, Arkwright and Kay had begun their revolution in the textile industry. Wool, like cotton, came out of the kitchen and into the mills.

The end of the eighteenth century saw such a speed-up in wool manufacture that local farmers were for the first time unable to cope with the demands of the industry. To keep the mills working, manufacturers were forced to look elsewhere for much of the raw wool that they needed. Spain and Germany helped to make up the deficit.

British Wool Industry

Throughout the nineteenth century, Yorkshire gradually became established as the centre of the world's wool trade, with other areas in Britain providing tweeds and special weaves. Industrial Britain, thriving in the prosperity that steam had brought, became the greatest textile manufacturing centre of the world.

As the demand for raw wool grew, manufacturers began recovering used wool from rags and old cloth. Fibres from these sources were respun and woven into cloth that was cheap enough to reach the pockets of the poorer classes. So the demand for wool increased still more.

Victoria's reign was a period of peace and plenty, with the Empire vibrant and flourishing as never before. Raw materials poured from the colonies and dominions into the busy factories of Britain, to be turned into manufactured goods for the markets of the world. Not least amongst these was wool. For the insatiable appetite of the Yorkshire mills was being fed by the great new wool-producing industries of the Dominions. Sheep in their millions were raised on the vast plains of South Africa, Australia and New Zealand, supplying thousands of tons of wool every year to the mills of Britain.

These dominion flocks were in the main bred from the famous merino sheep, which Spain had developed by careful breeding. The merinos yielded the finest and softest wool of all. In 1787, several merino rams and thirty-six ewes had been sent as a present to George III, but they did not thrive in Britain's damp climate.

South Africa

In the dominions the story was different. Two rams and four ewes presented by the King of Spain in 1789 to the Dutch Government became the nucleus of the great dominion flocks. Bred by Colonel Gordon, commander of the Dutch East India Company at the Cape, these merinos flourished in the warm climate of South Africa.

The colonization of South Africa during the nineteenth century was largely brought about by farmers seeking fresh pastures for their expanding flocks of sheep. By 1888, Cape Colony alone had 10½ million sheep; by 1927, South African flocks amounted to 44 million sheep with an annual wool clip valued at £18 million. During the 1930s, drought and depression reduced the numbers. But South Africa has rebuilt her flocks and produces a wool that is noted for its fineness and softness.

Australia

In Australia, similarly, wool production has become a major factor in the economy of the country. More than 134 million sheep graze on the Australian plains; they are direct descendents of 26 ewes and rams bought in 1795 from Colonel Gordon's merino stud at Cape Town.

As the number of sheep has increased so has their output of wool been multiplied by selective breeding. Modern merino flock sheep will yield 6.4kg (14 lb) of wool per head compared with 1.8kg (4 lb) of the early Spanish merinos.

Over three-quarters of Australia's sheep are merino breeds reared primarily for their wool. The clip is worth hundreds of millions of pounds a year.

New Zealand

New Zealand, the third great sheep farming country, has a climate more closely resembling that of Britain. The sheep are reared to a much greater extent for their mutton and lamb. Cold storage, developed in 1882, paved the way for the rapid growth of New Zealand's meat trade, and on the great Canterbury plains are bred the Southdown-Romney crosses that supply us with our Canterbury lamb.

But New Zealand is not solely concerned with meat. Dual purpose breeds of sheep now provide "Crossbred" wool in addition to meat. With more than 60 million sheep, New Zealand is a major wool producer.

World Production of Raw Wool (Greasy Basis) (1979–80)
(Million kg)

Argentina	176
Australia	722
Brazil	35
Bulgaria	35
Canada	1
Chile	20
France	24
Greece	10
India	35
Iran	39
Iraq	18
Irish Republic	9
Italy	12
Lesotho	2
Morocco	21
New Zealand	353
Pakistan	39
Peru	12
Portugal	14
Romania	37
South Africa	110
Soviet Union	472
Spain	29
Turkey	57
United Kingdom	51
United States	47
Uruguay	73
Yugoslavia	10
Other	248
Total	2,711

From these great countries, and from the U.S.A., South America and the U.S.S.R., comes a large part of the wool supplies of the world. Wool is an expensive fibre compared with cotton – it involves a second step, the animal, in its production. But wool is a quality fibre with a sustained demand from those who can afford it.

As the standard of living of the people throughout the world improves, so does the demand for wool increase. And as mass production brings down the price of wool garments, they are reaching into new fields.

Even in countries of the Far East, like China and Japan, where King Cotton has reigned unchallenged for centuries, the demand for wool is increasing. For wool will always sell on its merits in spite of its higher price. Even today, however, the consumption of wool in some countries is surprisingly small. New Zealand has a per capita consumption of up to 2.5kg (5.5 lb) a year, and that for the more prosperous European countries is 0.9–2.7kg (2–6 lb), but there are other European countries, like Poland, using less than a quarter of this amount. There are many millions of people, especially in Asia and Africa, whose means will not stretch so far as to allow them to possess a single woollen blanket.

The potential market for wool, therefore, is virtually unlimited. Only one thing restricts its use, and that is its limited production. Wool is so different a fibre, so unique and valuable in its properties, that were it cheaper it would be used for many purposes where at present we make do with other fibres.

Recovered Wool

The supply of raw wool available to the world every year amounts to about 2700 million kg (6000 million lb). After scouring, this is reduced to some 1550 million kg (3400 million lb) of pure wool.

This crop of wool is insufficient to meet the world's needs, and the supply is augmented to some extent by re-using wool which has already been made into yarns and fabrics, and even worn. Recovered wools of this sort are usually mixed with fleece wools and used for medium or low quality goods. Sometimes, fabrics are woven with cotton warp and a recovered-wool weft.

The rags and waste-fabrics used as raw materials for recovered wool are sorted and oiled before being opened out or teazed to fibre between rollers covered with wire teeth. There are three main types of recovered wool.

SHODDY is wool recovered from fabrics which have not been

excessively milled (i.e. felted) during manufacture. Cloths such as tweeds, knitted goods and worsteds provide the bulk of shoddy supplies. The wool fibres in the yarns of the fabric have not been matted together deliberately to give them a felted appearance, and they can be teazed apart with a minimum of damage.

MUNGO is made from cloths such as velours and meltons, which have been milled or felted during manufacture. The fibres in these cloths are in a more matted condition than they are in an unmilled cloth. They are more difficult to disentangle, and suffer more damage in the process.

EXTRACT consists of wool recovered from cotton/wool union fabrics. The cotton is removed by treating the fabrics with hydrogen chloride or dilute sulphuric acid; the wool that remains is teazed apart.

In general, recovered wools are poor in quality compared with fleece wools. The so-called 'rag-grinding' and (for yarn waste) 'garnetting' processes used for teazing the fibres apart tends to snap the fibres and to remove some of their surface scales. The short length of the fibres, together with the surface and other damage, cause a lack of firmness and poor handle in fabrics with a high content of recovered wool. Garments made from low-grade re-covered-wool fabrics do not wear as well as those from fleece wool, and tend to lose their shape. Moreover, recovered wools do not usually dye to such rich colours as new wool, so that recovered-wool garments are often dull in shade.

All Wool and Virgin Wool

A fabric or garment labelled as 'all wool' is not necessarily made from new fleece wool; it may contain a proportion of recovered wool. As such fabrics are inferior to those from new wools, it is customary to refer to new-wool materials as 'virgin wool'. The 'Woolmark', which designates such virgin wools, guarantees that a fabric is made from new wools.

PRODUCTION AND PROCESSING

When primitive man selected the sheep for domestication, he was guided in his choice by his clothing needs. He wanted an animal that would provide a skin of a size suitable for use as a human garment; and he wanted, at the same time, a creature that grew a soft and comfortable fleece. The sheep was an obvious choice.

The ancestors of our modern sheep grew coats of fibres that

served in two essential ways. On the outside of their fleece was a layer of long, coarse hairs which acted as a protective overcoat; these hairs were shed every spring. Under the layer of coarse hair fibres, the sheep grew an undercoat of finer hair, much more delicate and downy; this inner layer acted as a blanket to keep the animal warm. This insulating layer has given us the textile fibre that we now know as wool.

Modern sheep have been bred to provide as large a proportion of wool as possible. Sheep that are used primarily as a source of wool may carry only a trace of the outer covering in their fleece. Certain mountain breeds retain a relatively high proportion of coarse hair fibres. These fibres are sometimes unusually white in colour, and are opaque; in the finer woolled breeds they are generally regarded as a sign of poor breeding.

The merino, most important of all the sheep used as a source of wool, has almost entirely dispensed with its outer coat. Moulting no longer takes place, and the wool will go on growing year after year if it is not cut off.

Quality of Wool

Merino Wool

The raising of sheep for wool is now an important industry in many countries, and the quality of different wools is correspondingly diverse. The merino sheep, which produces fine, soft wool forms the basis of wool production in Australia, South Africa and South America. Immense flocks of merinos are raised in these countries.

Australia

The quality of the merino wool from these sources depends upon environmental conditions, and upon the hereditary characteristics of the sheep. Port Philip wool is reputed to be the finest Australian fibre, and is used for making high-quality woollen and worsted fabrics. Sydney and Adelaide wools are not quite so fine as Port Philip, and they are a shade yellower in colour. Tasmanian wool is of first-rate quality and washes a beautiful white.

South Africa

Wool from South Africa is very crimped or wavy, and has a good white colour after washing. It is used for good-quality worsteds and woollens.

South America

South American wool is not generally of such good quality as wool from Australia or South Africa. Much of the South American wool is used by continental manufacturers. The best quality South American wool comes from Montevideo, with Buenos Aires next. Punta Arenas wool, which comes from cross-bred sheep, is a bulky wool and is widely used in making hosiery.

Europe

In Germany, France, Spain and other European countries, and in the U.S., the merino has been reared successfully and often provides wool of high quality. Saxony and Silesian merino wools, for example, have the reputation of being the finest in the world. The French Rambouillet is renowned for its high-quality wool.

Crossbred Wool

The merino sheep imported into Britain by George IV did not find conditions to their liking. In some other countries, however, crossing of the merino with other breeds of sheep was highly successful, and breeds originating from such crosses are now of great importance as wool and meat producers. Sections of the New Zealand sheep industry, for example, are based on cross-bred sheep which provide both wool and mutton. Australia and South Africa also export a great deal of wool produced by cross-bred sheep.

British Wool

British-grown wool can be graded into four main types described as lustre, demi-lustre, down and mountain wools.

LUSTRE WOOL comes from Lincoln, Romney Marsh, Cotswold and Leicester sheep. It is up to 30.5cm (12 in) long and is made into lustrous dress fabrics, buntings and linings.

DEMI-LUSTRE WOOL is shorter and has less pronounced lustre. It is made into serges, dress fabrics and curtains.

DOWN WOOLS are medium length 75–100mm (3–4 in) staple; fibre is curly and has a crisp handle. Sussex or Southdown sheep provide some of the finest of English wool; it is made into hosiery, cheviot suitings and flannels.

MOUNTAIN WOOLS from breeds such as the Scottish Blackface vary greatly in quality and length. Blackface wool is long and coarse and contains a high proportion of kemp fibres.* It is made into tweeds and carpets; the famous Harris tweed is commonly made from Blackface wool.

Cheviot, a medium length wool, is made into knitwear, tweeds, worsteds and cheviot suitings. It is strong and of bright colour, and has good felting properties. Welsh wool is used largely for making flannels.

Irish wool is too thick to be spun into fine yarn, and is used for homespun tweeds and woollens and for carpets.

In the Shetlands, wool is combed from the sheep instead of being removed by clipping. It has a very soft handle although it may contain a high proportion of hair, and is knitted into the shawls, cardigans and other garments that are known the world over.

Asian Wool

In China and other parts of Asia, in Turkey and Siberia, the production of wool is of growing importance. The wool is often long and coarse compared with fibre produced in Australia, South Africa and the other great wool-producing countries.

As in the case of cotton and other plant fibres, the quality of wool depends greatly upon the conditions under which it is grown. Wool derives from a living creature, and it is affected not only by the hereditary characteristics of the sheep but by the environment in which the sheep has lived.

Wool Production

Wool fibres grow from small sacs or follicles in the skin of the sheep. The wool fibres grow in groups of 5–80 hairs and there are 1550–3410 per sq cm (10,000–22,000 per sq in).

A typical Hampshire sheep will have some 16–40 million fibres in its fleece; a Rambouillet between 29 and 97 million; and an Australian merino may carry as many as 120 million individual wool fibres. These fibres grow on the average at the rate of 2.5cm (1 in) in two months; altogether they represent a considerable drain on the resources of the animal.

Shearing

Sheep are normally shorn of their fleece every year (in some countries, e.g. South Africa, up to twice a year). On large stations,

* 'Kemps' are a short wavy type of hairy fibre which are shed periodically.

the fleece is removed in one piece by power-operated clippers. In efficient hands, the sheep is parted from its wool in two and a half minutes. A first-class shearer will get through two hundred sheep a day, from which he will clip a tonne or more of wool. This type of wool is known as 'fleece or clip wool'.

Immediately after it has been removed, the fleece is 'skirted'. This involves pulling away the soiled wool around the edges. Then the whole fleeces are graded by experts who judge the fineness, length, colour and other characteristics. Finally, the various grades are packed into large sacks and then sewn up into the bales which are a familiar sight in any Yorkshire town. Each bale contains about 136kg (300 lb) or more of wool.

Slipe Wool; Mazamet

Wool is also removed from the pelts of slaughtered sheep. The pelts are treated with lime and sodium sulphide or some other depilatory. This loosens the wool, which can be pulled away without damaging the hide. This wool is called 'slipe wool'. Hides can also be subjected to bacterial action to loosen the wool; the product is called fellmongered or 'Mazamet' wool, from the French town of Mazamet where it is produced in quantity. Wool removed from the skin in these ways is usually inferior in quality to clip wool; it is often used mixed with fleece wool.

Wool Sales

In the U.S.A. and in some South American and other countries, wool is sold direct by the shearer by private treaty, often with a minimum of preparation. In Australia, New Zealand and South Africa, fleeces are skirted and classed carefully into recognised categories before being sold at public auction. In Great Britain wool is sold by farmers to the British Wool Marketing Board who bulk-class the fleeces before selling the wool at public auction.

British Wool Industry

Raw wool reaching Britain is sent off to the great manufacturing centres. Most important of all is the West Riding of Yorkshire, where woollen fabrics of almost every type are made. Heavy and medium class woollens come from the mills of Leeds, Dewsbury, Morley and Batley. Carpets are made in south west England, Ireland and elsewhere.

The finest saxony woollens are made in the West of England, near Trowbridge. Tweeds and cheviots come from Galashiels, Hawick and other border towns, where Scottish cheviot wool is spun and woven into the famous 'Scotch Tweeds'. From Aberdeen come the world-renowned Crombie materials. Witney makes blankets, and Rochdale makes flannels. In the Leicester and Nottingham districts, woollen yarn is made into knitted materials.

The worsted industry is less scattered, and is centred in Yorkshire. Bradford – 'Worstedopolis' – makes all manner of dress goods and linings. Huddersfield has a reputation for the finest quality worsted suiting materials. Wakefield, Keighley and Halifax make worsteds in a great variety of forms.

Preparation of Wool for Spinning

Grading and Sorting

The quality of wool varies greatly with the breed of sheep and the conditions under which it has lived. Quality depends also upon the characteristics of the individual sheep, and upon the region of the sheep's body in which the wool has grown.

As the fineness and length, the softness and colour of wool determine the uses to which the fibre is put, grading and sorting of wool are essential preliminaries to spinning.

Counts

The first grading of wool takes place as it is being prepared for sale. The price of raw wool depends upon the buyer's assessment of its fineness and length. Quality is defined by numbers which at one time described the limiting fineness or 'count' of yarn into which it could be spun on the English worsted count system. An 80s wool, for example, was at that time considered capable of being spun into a yarn of 80s count. This meant that 1 lb (454g) of wool would yield 80 hanks each containing a fixed length of yarn. In the worsted industry, the standard length of a hank is 560 yd (504m). 1 lb (454g) of 80s wool would thus be capable of spinning into 80 hanks each of 560 yards if it was spun to the finest limit. Under modern conditions these spinning limits are no longer valid, but the traditional numbers continue to be used to describe wool quality or fineness. (In woollen manufacture, the unit of skein-length is shorter and varies from place to place. In Yorkshire, the standard is 256 yd (230.4m), whereas in the West of England it is 320 yd (288m).

Thus, although wool quality numbers no longer relate directly to

91

the worsted yarn count system, they form an arbitrary scale whereby the higher the quality number, the finer is the wool. A merino wool usually lies between 60s and 100s; wool from cross-breds is 36s – 60s, and coarse wool such as that used in carpets is less than 44s.

Fibre Length

The average length of wool fibres is described by special terms such as 'combing' or 'clothing'. When the fibres are long enough to undergo combing and be made into worsteds –65mm (2½ in) or more – they are combing wools; 176mm (7 in) they are long wools. Fine fibres, 38–65mm (1½–2 in), can be combed in the French comb and are 'French combings'. Short wools of less than about 32mm (1¼ in) are described as 'carding or clothing wools'.

Classifying

When the bales of wool are opened in the mill, the fleeces are skirted if this has not already been done. The fleece may be classified as a whole or, if variable in quality, separated into sections such as shoulders, sides, back, thighs and britch and belly. In general, the shoulders provide the best wool, and the flanks a slightly lower quality wool. The belly, tail and legs yield the poorest quality of wool. In medium and long wool breeds of sheep, the head, legs and britch usually produce the highest proportion of hair and kemps. Some breeds, such as the Scotch blackface, produce hair and kemp fibres in all parts of their fleece.

Lamb's Wool

The finest wool is obtained from young sheep. Lamb's wool clipped at eight months is very fine and of excellent quality.

Hog Wool

Six months later, when the sheep is fourteen months old, the wool ('hog') is stronger and thicker. As the animal grows older, the quality of the wool decreases slightly.

Altogether, a sheep, on the average, will provide between 0.9 and 4.54kg (2–10 lb) of wool per year, according to breed.

Scouring

Raw wool is dirty and contaminated with natural substances that must be removed before processing can be carried out. Often, as much as 50 per cent of the weight of raw wool consists of impurities

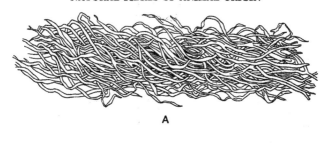

A

B

Wool. Woollen and Worsted Yarns. In a woollen yarn (A), the random arrangement of the fibres results in a bulky yarn with a fuzzy surface.

In a worsted yarn (B), the fibres are lying more parallel, and are more tightly twisted, producing a thinner yarn with a smoother surface.

---

of one sort or another. In addition to dust and dirt, the wool is mixed with natural grease (yolk) and with dried perspiration (suint). In general, the finer wools such as merino contain a higher proportion of natural impurities than the coarser wools.

Raw wool is washed or scoured by being agitated gently in tanks filled with warm water containing detergent. The raw wool is propelled gently through a tank, then squeezed between rollers and carried into another tank. It may pass through four or more tanks in this way, until eventually it is rinsed in clean water. It is then dried until about 20 per cent of water remains in it.

Other methods are now being used for cleaning raw wool. It can be washed, for example, in a grease solvent fluid.

Although these processes clean raw wool by removing unwanted substances from it, they do not necessarily produce a white fibre. The wool may contain natural colouring matters.

Processing and Spinning

Wool is spun into two types of yarn, woollen and worsted, and the treatment of the fibre after scouring varies accordingly.

Woollen yarns are thick and full; the fibres are held loosely and subjected to only a limited twist during spinning. These yarns, made usually from short-staple wool, are woven into thick, full-bodied materials such as tweeds or blankets and used for knitting.

Worsted yarns, on the other hand, are finer, smoother and firmer. The fibres in worsted yarns are aligned so that they lie closely alongside each other in the direction of the yarn; for woven fabrics they are twisted tightly together to form a fine strong yarn. Worsteds are spun commonly from fibres 5–38cm (2–15 in) long. These yarns are woven into fine dress materials and suitings. Worsted spun yarns with less twist are used for knitting yarns and knitted fabrics.

Woollen Yarns

Although scouring will wash most of the grease and suint from the raw wool, it does not necessarily remove any burrs, twigs and other vegetable material from the fibres. These impurities can be destroyed by steeping the wool in dilute sulphuric acid and then heating at high temperature. The cellulosic material is charred and can then be broken up and beaten out of the wool. This process is called CARBONIZING.

Burrs and other vegetable impurities can also be removed mechanically by passing the wool through heavy crushing rollers before the intermediate stage of carding, so that the vegetable impurities are removed as a powder in subsequent carding.

Often, woollen yarns are spun from a mixture of new wool with reclaimed wool, or with rayon, cotton or other fibres. After scouring and cleaning, the wool is blended to mix various grades together and to incorporate any other fibres.

Carding

The wool is now ready for carding in machines equipped with rollers covered with sharp steel wires which disentangle the matted fibres. There are usually three parts to the machine used in woollen carding, called the scribbler, intermediate and carder. They all perform in much the same way, separating the fibres and mixing them thoroughly. The wool emerges from the carding machine as a thin blanket of fibres about 1.5m (5 ft) wide, holding together as a fluffy mass.

Woollen Spinning

This blanket of loosely-held fibres is split up and formed into ribbons which hang together sufficiently to be able to support their own weight. The fibres in the ribbon – or condensed slubbing, as it is called – are lying quite higgledy-piggledy. The slubbings are drawn out and spun to form a yarn. The fibres in the slubbing have been lying in all directions, and when they are twisted during spinning they produce the soft yarn so characteristic of woollen goods.

Woollens spun from merino wool are described as *saxony woollens*; woollens made from cross-bred wools are generally called *cheviot woollens*.

Worsted Yarns

Wool to be spun into woollen yarns may be scoured and dried by commission firms and stored until wanted. It is bought as blends or separate types. But wool for spinning into worsteds is normally scoured and dried immediately before carding.

After drying, the wool is carried straight to the carding machines. The mass of fibre is opened, teased and cleaned, emerging in the usual flimsy sheet. This is brought together to form card slivers which are wound up into balls or allowed to fall into deep cans.

Gilling

When the wool is of very long staple, for example 23–38cm (9–15 in), instead of being carded it is often subjected to a process called gilling. The wool is passed through a gilling machine in which the moving strand is combed by rows of pins that move rapidly through the mass of fibres. The fibres are straightened and aligned in the direction in which the wool is moving.

Combing

This alignment of the wool fibres so that they lie parallel is a characteristic feature of worsted manufacture. The wool, after carding or 'preparing', is passed through a combing machine which continues the process, combing out short fibres and aligning the long fibres accurately alongside each other. It is also usual to pass the wool through gilling machines between carding and combing and after combing.

The combed wool in this form of an untwisted strand called the 'top', is then drawn out into a roving and spun by twisting it into a yarn. The long fibres, lying parallel to one another in the combed roving, are able to cling tightly together on twisting to form a fine, strong worsted yarn.

Worsted Spinning

The spinning of worsted yarns is carried out in one of four different ways.

FLYER SPINNING is used with long wools (100–250mm (4–10 in), mainly 150mm (6 in) fibres, yielding smooth yarns used largely for hosiery.

CAP SPINNING (almost obsolete) is used very largely with botany and fine cross-bred wool. Its production rate is high and the yarn tends to be hairy.

RING SPINNING has the highest output and is used for making the fine yarns. It now accounts for 80% of the world's spindles.

MULE SPINNING (now obsolete) is used for making continental worsteds, giving a full, soft yarn.

In worsted spinning, 454g (1 lb) of wool may spin into more than 63,000m (70,000 yd) of yarn which is made into hard-wearing, high-quality fabrics. The use of expensive raw wools and the number of processes involved in manufacture mean that worsted fabrics are usually high-priced.

As in the case of cotton, the finer wool fibres provide the better quality yarns and fabrics. Fine fibres can be spun into uniform, smooth yarns that yield firm, lightweight materials which are soft and warm, drape well and retain their shape when made into garments.

The long, fine fibres of good quality wool have many more scales to the cm than the coarser fibres, and are usually more crimped.

Worsteds made from merino wool are described as *botany worsteds*; worsteds made from cross-bred wools are *cross-bred worsteds*.

Twist

The amount of twist put into a yarn when it is spun depends upon the nature of the fibres and purpose for which the yarn is designed. Long fibres need less twist to hold them strongly together than short ones do. Thick yarns need less twist than finer ones. In general, worsted weaving yarns are spun much tighter than woollen yarns.

When the yarn is to be made into a hard-wearing fabric, such as a worsted for suiting or a gabardine for an overcoat, it is twisted tightly to give a strong, weather-resisting material with a comparatively hard texture.

If, however, the yarn is to be used for clothes that are worn near the skin, it must be soft and yielding, warm and light. The yarn is therefore given less twist than is needed for a worsted.

Fabrics from lightly-twisted yarns, unless subjected to modern finishing treatment, e.g. 'Superwash', will also felt more easily than the tighter-spun yarns. Low-twist yarns are used in fabrics that are to be milled or felted deliberately, for example in making meltons.

Bleaching

If the wool is to be dyed to especially bright or pale shades, or when it is to be spun and woven into white yarns and fabrics, it can be whitened by bleaching. This is usually carried out on the yarn or fabric, rather than on the loose wool.

STOVING consists of the exposure of wool to sulphur dioxide in a large closed chamber. The gas, produced by burning sulphur, converts coloured impurities in the fibre into colourless substances. Wool is usually bleached in this way in the form of skeins of yarn or as fabric. The material is left overnight in the chamber, and is well washed the following morning.

Although stoving produces a good white wool, the effect is often impermanent. The wool tends to revert to its natural colour as oxygen in the air converts the colourless substances into coloured materials again.

HYDROGEN PEROXIDE BLEACHING is a more expensive process than bleaching by sulphur dioxide. The effects, however, are more permanent. The wool is immersed for several hours in a dilute solution of warm hydrogen peroxide. The coloured impurities are destroyed and the effects of the bleaching are permanent.

Dyeing

Wool may be dyed at several stages during manufacture, from raw wool to piece goods. The fibre passes through many processes when it is being made into textiles, and dyes must be chosen which can withstand the conditions they will have to meet. In addition, the dye must, of course, be satisfactory with respect to the purposes for which the finished fabric will be used.

THE WOOL

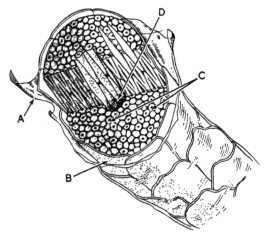

The Wool Fibre. In a typical wool fibre there are four distinct regions, which can be distinguished readily through the microscope.

(A) *The Outer Sheath or Epicuticle*. This is the outermost layer, and is a thin, water-repellent membrane. It is the only non-protein part of the fibre, and it protects the fibre like a covering of wax.

The membrane repels water, and in this sense acts as a waterproof coating on the fibre. Wool fabrics will repel the water that falls on them during a short rain shower. The epicuticle is, however, permeated

98

FIBRE

by many microscopic pores, through which water vapour may penetrate into the interior of the fibre. Wool fabrics will thus absorb water vapour from the body without feeling damp, and will release it again slowly into the air.

The outer membrane is easily damaged by mechanical treatment, and the wool then becomes more easily wetted.

(B) *The Scale-Cell Layer.* Beneath the epicuticle there is a layer of flat, scale-like cells which overlap like the shingles on a roof. The free ends of the scales point towards the tip of the fibre. The epicuticle and scaly layer together form the cuticle of the fibre.

(C) *The Cortex.* Enclosed within the cuticle is the bulk (more than 90 per cent) of the wool fibre. This is the cortex, which consists of millions of long spindle-shaped cells, thick in the centre and tapering to points at each end. These cortical cells are 100–200 microns in length and 2–5 microns wide.

The cortical cells are themselves built up from fibrous components called fibrils, which are in turn constructed from protofibrils. These may be seen through the electron microscope as globular particles which possibly consist of the keratin molecules themselves.

The cortex of the wool fibre is constructed in the form of two distinct sections in which the proteins differ slightly in chemical and physical properties. The fibre can be regarded, in fact, as being formed from two half-fibres of semi-circular cross section which are joined lengthwise. The sections are twisted spirally round one another, the twists being in phase with the natural waviness of the fibre.

(D) *The Medulla.* Many wool fibres, especially the coarser fibres, have a hollow space running lengthwise through the centre. This is the medulla. It may be empty, or may contain a loose network of open cell walls – *After The Wool Bureau Inc.*

A wide range of wool dyestuffs is now available, which enable the manufacturer to meet his most exacting requirements. Most important are the acid, chrome, premetalized, mordant, vat, reactive and milling dyestuffs.

Fine Structure and Appearance

The wool fibre grows from a sac-like organ, called the follicle, in the skin of the sheep. The mouth of the follicle forms a tiny hole in the animal's skin, and the wool fibre grows up through it from a growing point at the base of the sac.

The young wool fibre has a tip tapering to a point, but a fibre that has been cut retains the flat tip left by the shears. Above the skin level the fibre is dead material, mainly keratin, a protein similar to that from which horn, finger-nails and feathers are made. There is no way in which the wool can change its physical form once it has left the follicle; if the tiny rod of keratin is cut, it remains cut.

Although the wool fibre is essentially a rod of protein, it is constructed in the first place as a result of living processes. The fibre is not merely a homogeneous rod, but consists of a complex structure built on a cellular basis.

There are four distinct regions in the wool fibre. On the outside is a delicate membrane forming the outer sheath or epicuticle. Beneath this is a layer of scale-like cells, followed by the cortex which makes up the bulk of the fibre. In the centre, there is commonly a hollow core or medulla.

Outer Sheath or Epicuticle

The outer membrane, the epicuticle, is about 100 Å thick. It repels water, but is permeable to water vapour which passes through microscopic pores in the sheath.

Scale-Cell Layer (Epithelial Scales)

This layer, together with the epicuticle, forms the cuticle of the fibre. It consists of horny, irregular scales, called epithelial scales, which cover the fibre. The scales project towards the tip of the fibre and overlap like the tiles on a house-roof. In the finer fibres, they encircle the fibre completely, giving it the appearance of a stack of flowerpots sitting one inside the other. In coarser wools the scales may not encircle the fibre, but overlap in two directions.

These epithelial scales are, on average, $0.05-0.5\mu$ thick. They fit closely together forming a protective covering that plays an important part in controlling the properties of woollen yarns and fabrics.

The number of scales varies greatly, depending on the fineness of the fibre. The finest fibres, such as those in a saxony wool (merino), may have 790 to the cm (2000 to the inch). A coarse wool, such as that from a Scottish blackface, may have only 276/cm (700/in).

In fine wools, only 10μ or less of each scale may be exposed, although the scale itself is about 30μ long and 36μ wide. In coarser fibres, as much as 20μ of the scales may be exposed although they do not project so far as those of the finer fibre.

Cortex

The cortex forms the main central portion of fine wool fibres. It is built up from long, spindle-shaped cells which provide the strength and elasticity of the wool fibre. These cortical cells are held together by a strong binding material.

The average length of the cortical cells is about $80-110\mu$, the average width $2-5\mu$ and the thickness $1\cdot2-2\cdot6\mu$.

The remains of a nucleus is usually visible in each cell, and the cells are sometimes coloured by natural pigments. They have a striated appearance.

Detailed studies of the cortical cells of wool have shown that they are built of many tiny fibrils, which are in turn constructed from even smaller micro-filaments. These elongated structures lie roughly parallel to the long axis of the spindle cells.

There are two types of cortex cell – orthocortex and paracortex – which differ in the way their constituent fibrils are arranged.

In crimpy merino wools the orthocortex and paracortex cells form two half-cylinders lying alongside one another. The two half-cylinders are twisted spirally along the length of the fibre, with the softer orthocortex cells forming the outer edges of the crimp curves. (See fig., page 103.)

In coarser, straighter wools, such as lustre wools, the orthocortex cells form a central cylinder with the paracortex cells forming a sheath around them.

Medulla

The medulla of the wool fibre is sometimes a hollow canal, and in coarser fibres may consist of a hollow tubular network. Coarse and medium wools are characterized by the presence of a greater proportion of medullated fibres. In the majority of merino fibres, the medulla is either absent or so fine as to be almost invisible. It may account for 90 per cent of a kemp fibre.

The coarse hair-fibres, with their pronounced medullae, are usually straighter and more lustrous than the finer wool fibres. They have poor spinning properties and cause difficulties by dyeing to a lighter shade than the true wool fibres. This is a result of the internal diffusion of light by the cells of the medulla and the fact that there is not much thickness of cortex to take up the dye.

Dimensions

The dimensions of wool fibres vary between considerable limits. Fine wools are about 38–125mm (1½–5 in), medium wools 65–150mm (2½–6 in) and long wools 125–375mm (5–15 in). The measurement of the actual length of a wool fibre is complicated by its crimp (see below). The stretch length of a fibre may be nearly twice that of its natural length.

The average width of a top quality merino fibre is about 17μ; a medium wool fibre is about 24–34μ, and a long wool about 40μ. The fibres are therefore thicker, on the whole, than cotton fibres. Wool fibres are roughly oval in cross-section.

Crimp

Wool fibres are unique among natural textile fibres in having a wavy structure, which is of supreme practical importance. It enables the fibres to hold together when twisted into a yarn, just as the convolutions of cotton fibres do in a cotton yarn. This waviness or crimp of wool fibres is most pronounced in the fine wool fibres. The best merino wools, for example, will have as many as 12 waves to the cm (30 per inch) in a fibre 15μ wide. Lower quality wool will have 2 or even less to the cm (5 per inch).

This waviness endows wool fibres with unusual elasticity; they can be stretched out like small springs and will return to their wavy form when released. This ability to return to the natural form is made possible by the inherent elasticity of the material of the fibre itself.

Lustre

Wool fibres have a natural lustre which varies in its characteristics, depending on the type of wool.

Lustre seems to depend very largely on the nature of the fibre surface. Light is reflected from the fibres in such a way as to create a lustrous appearance. The 'lustre wools' from Lincolns, Leicesters

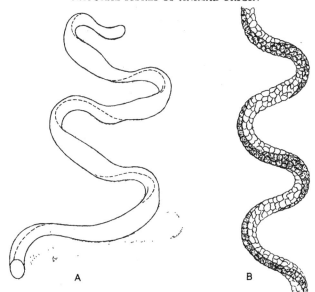

A   B

Crimp. The wool fibre has a natural waviness or crimp, which is unique among natural fibres of major commercial importance. The crimp does not consist of waviness in a single plane, but takes the form of a three-dimensional waviness as shown above. It is related to the spiral form of the two core sections of different constitution, which twist spirally around one another in phase with the twists of the crimp – *After The Wool Bureau Inc.*

---

and other English breeds have a surface which provides a silky lustre.

Merino wools do not reflect light so perfectly. They have a delicate lustre.

Colour

Most of the wool from modern sheep is white or near-white in colour. Some breeds of sheep produce a quantity of brown or black wools, the proportion being highest in the breeds that provide the coarsest wool.

Tensile Strength

Wool has a tenacity of 8.8–15.0 cN/tex (1.0–1.7 g/den) dry, and 7–14 cN/tex (0.8–1.6 g/den) wet.

The tensile strength is 1190–2030 kg/cm² (17,000–29,000 lb/in²).

Elongation

Wool has an elongation at break of 25–35 per cent under standard conditions, and of 25–50 per cent when wet.

Elastic Properties

Wool is an unusually resilient fibre. Its high elongation at break is combined with a high elastic recovery that gives wool unique properties in this respect. The natural crimp of the wool fibre contributes to the overall elasticity, but the property is a fundamental one which derives from the curled, cross-linked structure of the wool molecules themselves.

Wool has an elastic recovery of 99 per cent at 2 per cent extension, and of 63 per cent at 20 per cent extension.

Specific Gravity

Wool is a light-weight fibre of specific gravity 1·32.

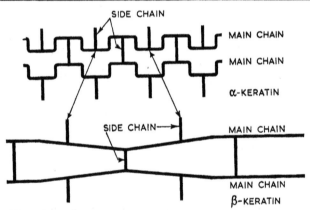

Elasticity of Wool. Wool fibres are highly elastic, and after stretching will return to their original shape again. In this respect they have a resemblance to rubber, and the reason for this behaviour is in both cases the same. The molecules of wool, like those of rubber, are highly folded.

In the wool fibre, millions of protein molecules are lying alongside one another, held together at intervals by chemical cross links. When the fibre is stretched, the chains unfold; then, when the stretching force is removed, they return to their folded state again. The simplistic diagram shows the folded molecules of alpha-keratin changing into the stretched molecules of beta-keratin – *After The Wool Bureau Inc.*

Effect of Moisture

Wool absorbs moisture to a greater extent than any other fibre, and yields it up readily to the atmosphere. Under ordinary atmospheric conditions, wool will hold 16–18 per cent of its weight of moisture. Under suitable circumstances, wool will absorb about a third of its weight of water.

The readiness with which wool adjusts its water-content in response to changes in atmospheric conditions is of great commercial importance. The weight of wool changes accordingly, and the definition of moisture conditions form part of any contract in the buying and selling of wool.

When a fibre absorbs moisture in this way, heat is liberated. This effect is particularly marked in the case of wool, and a woollen garment will become warmer as it absorbs moisture from the air.

In common with all other fibres, wool is composed of molecules that are long and thread-like. The atoms forming the molecule of wool keratin are joined together in such a way as to form molecular threads. The ultimate particles of wool material are, in fact, like fibres in themselves. It is the association of millions of these molecules to form the bulk of the wool keratin that gives wool its fibre characteristics. The molecules are aligned one beside the other, held together by chemical and electrical forces.

As in the case of cellulose, the mutual forces of attraction between the keratin molecules of wool can be influenced by the absorption of water. When wool absorbs water, the relatively small molecules of water are able to penetrate between the long molecular threads, prising them apart and so weakening their hold one upon the other.

In the case of wool, the forces that play between neighbouring keratin molecules are more complex than those which hold cellulose molecules together in cotton or flax. Keratin molecules are bridged by definite chemical bonds which are not easily broken. The action of water on wool is complicated by this chemical cross-linking that holds wool molecules together.

When wool is soaked in water at ordinary temperature, it absorbs water and swells until its volume increases by about one-tenth. As the wool dries, it returns to its original size.

Hot water or steam have a deleterious effect on wool over a period of time. The wool loses its strength; heated under pressure at 120°C. it will eventually dissolve.

In boiling water, wool fibres lose much of their resiliency; they become plastic. The plasticizing effect of water increases with increasing temperature and time. The longer the wool is heated, and the higher the temperature, the greater the effect of water will be.

Effect of Heat

Wool becomes weak and loses its softness when heated at the temperature of boiling water for long periods of time. At 130°C. it decomposes and turns yellow, and it chars at 300°C. As it decomposes, wool gives off a characteristic smell, similar to that from burning feathers.

Wool does not continue to burn when it is removed from a flame. Each fibre forms a charred black knob; this is a test used in the identification of wool.

Effect of Age

Wool shows little deterioration when stored carefully.

Effect of Sunlight

The keratin of wool decomposes under the action of sunlight, a process which begins before the wool has been removed from the sheep. The sulphur in wool is converted into sulphuric acid; the fibre becomes discoloured and develops a harsh feel. It loses strength and the dyeing properties are affected.

Wool subjected to strong sunlight is particularly sensitive to alkalis, including soapy water.

Chemical Properties

Whereas cotton, flax and the other plant fibres are basically cellulose, wool is protein. Keratin, the substance of wool, is similar in its essential structure to the other proteins from which much of the animal body is built. The chemical structure of wool differs only slightly from that of feathers, hair and horn.

In their chemical behaviour, proteins are quite different from cellulose. They are more easily degraded and attacked by chemicals, particularly of certain types. They do not, in general, have the resistance to environmental conditions that is so characteristic of cellulose.

Contaminants

As it comes from the sheep, wool is contaminated by a variety of materials. The animal itself exudes a supply of wool grease and suint onto the fibres as they grow. Raw wool may contain 20 per cent or more of grease, and 12 per cent of suint. The finer wools contain the highest proportion of these natural impurities.

WOOL GREASE is removed from the wool before spinning, and is a commercially valuable material. In its purified form it is known as lanolin, which is widely used as an emollient.

Lanolin itself has become a source of medicinally important chemicals such as cholesterol.

SUINT consists essentially of salts that are left behind when the perspiration exuded by the sheep has dried. It contains a high proportion of potassium salts.

Keratin

The wool protein itself, keratin, is, like all proteins, an extremely complex chemical. It contains the elements carbon, hydrogen, oxygen, nitrogen and sulphur.

The general sensitivity of wool keratin to chemicals affects all aspects of the processing of wool. Bleaching, for example, must be carried out with great care. Hydrogen peroxide, commonly used as a bleach for wool, will damage the fibre if the conditions of bleaching are not adequately controlled.

Wool is discoloured and damaged by alkaline hypochlorite solutions; many domestic bleaches based on hypochlorite should never be used on wool.

Effect of Acids

Wool is attacked by hot concentrated sulphuric acid and decomposes completely. It is in general resistant to other mineral acids of all strengths, even at high temperature, though nitric acid tends to cause damage by oxidation. Dilute acids are used for removing cotton from mixtures of the two fibres; sulphuric acid is used to remove vegetable matter in the carbonizing process.

Effect of Alkalis

The chemical nature of wool keratin is such that it is particularly sensitive to alkaline substances. Wool will dissolve in caustic soda solutions that would have little effect on cotton. The scouring and processing of wool is carried out under conditions of low alkalinity. Even weakly alkaline substances such as soap or soda are used with care. Soda will tender wool and turn it yellow if used in too concentrated a solution, particularly if the solution is too hot. Ammonium carbonate, borax and sodium phosphate are mild alkalis that have a minimum effect on wool. Ammonia, carefully used, will not cause damage.

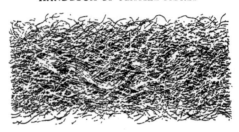

The Warmth of Wool. The natural crimp of the wool fibre results in a bulky or lofty fabric, each fibre standing away from its neighbours. Air is trapped in the spaces between the fibres, creating an insulating layer which restricts the passage of heat through the fabric. Depending upon the texture and thickness of the fabric, as much as 60 to 80 per cent of the volume of a wool cloth may be air.

The elasticity of wool enables it to retain this loftiness throughout the life of the fabric, the texture being restored after the fabric has been flattened.

The drawing above shows an edge-view of a lofty wool fabric.

Effect of Organic Solvents

Wool has a good resistance to dry-cleaning and other common solvents.

Insects

Wool is attacked by moth-grubs and by other insects. (See 'Wool in Use').

Micro-organisms

Wool has a poor resistance to mildews and bacteria and it is not advisable to leave wool for too long in a damp condition.

WOOL IN USE

One of the first things we meet on entering the world is wool. And, although it is no longer compulsory by law, wool is still very often our closest companion when we leave. Woollies are worn by babies because they are warm and airy. Wool clothes are healthy and hard wearing. They have many other properties – some desirable, some not. All are in one way or another a direct consequence of the properties of the fibre itself.

The wool fibre has excellent spinning characteristics. Its crimp enables the fibres to cling tenaciously together when they are spun.

Because of this, it is possible to make a relatively strong yarn from wool fibres without twisting them very tightly. Knitting wool, for example, can be spun very loosely and yet is quite coherent.

The crinkliness and resilience of the wool fibre and the looseness of the yarn are largely responsible for wool's warmth as a material. The fibre itself is a poor conductor of heat, but the real insulation is provided by the air trapped inside the fabric. Wool's high natural crimp means that woollen yarns and fabrics tend to be lofty open structures which hold much heat-insulating air inside them. The high elasticity and resilience of wool prevent the fibres from bedding down and the fabric becoming thin during wear.

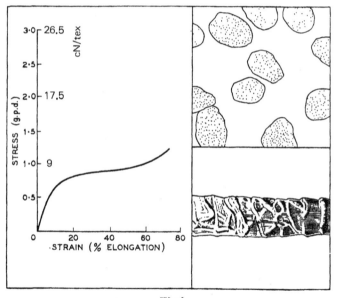

Wool

The flexibility and elasticity of wool fibres contribute to the high resistance of wool fabrics to crushing and creasing. The fibres can be distorted, but they will tend to return to their original shape again. Wool fibres can be bent backwards and forwards tens of thousands of times without breaking. Elbows and knees of wool suitings can withstand repeated flexing to which they are subjected.

Effect of Water

Wool's thirst for water is one of its most important characteristics. Water molecules can penetrate between the long molecules of keratin in the wool fibre, loosening the mutual grip of the molecules lying close alongside each other. The molecules are able to move more easily relative to one another, and the wet wool becomes softer and more plastic.

Setting

This lubricating effect of water on wool is only one aspect of the influence of water on the properties of the fibre. Water will attack wool keratin, causing changes in the chemical structure of the protein itself.

When a wool fibre is stretched, the coiled-up molecules are extended into a less-folded form. The extent to which this happens is controlled by chemical bonds and forces of attraction that link the long molecules laterally together. When the stretching force is removed, these lateral links pull back the deformed molecules into their original folded state.

If, however, links between the molecules are destroyed as the wool is held in its extended state, and new links are created between the molecules in their new positions, these new links will serve to hold the wool in its extended state. The fibre will have acquired a 'permanent' set.

This is the basis of practical techniques developed for setting woollen fabrics and garments into shapes that will be retained during normal use. The reactions involved in breaking down and rebuilding lateral links between wool molecules are almost invariably reactions in which water plays a part, and are accelerated by the use of elevated temperatures.

Ironing

In ironing a woollen garment, we remove the creases and set the fabric in the shape we want by subjecting it to a combination of moisture and heat. Creases are put into worsted trousers and pleated skirts by 'steam setting'; the fibres are induced to remain in their 'unnatural' shape by the combination of heat, moisture and pressure.

Absorption

Apart from the chemical effect of water on the wool fibre, the purely physical absorption of moisture is of great practical importance. The moisture disappears into the fibres and there is no feeling of clamminess or dampness about the garment, even though it may be holding quite a large quantity of water.

As it absorbs moisture in this way, a woollen garment will generate heat. Putting water onto wool is like adding it to quicklime; heat is produced, but not, of course, to the same degree. A woollen overcoat can therefore have a double warmth-inducing effect. On the one hand, the wool prevents heat escaping from the body and into the surrounding air. In addition, as it absorbs body moisture or rainwater, the wool generates heat on its own account.

When it has absorbed water, wool will get rid of it again slowly into the surrounding air. Evaporation of a liquid in this way always causes a drop in temperature; if this happens rapidly it can produce quite a chill. Wool, however, evaporates its moisture slowly and gently; there is no sudden cooling to chill the body surface and encourage rheumatism and similar ailments. Woollen bathing costumes, for this reason, are safe and healthy. They can be worn as they dry without chilling the body by permitting rapid evaporation of the water from the fabric.

Attack by Insects

This combination of unusual and valuable characteristics makes wool extremely popular as a high-quality fibre. Unfortunately, this popularity extends beyond the human race; wool is in great demand in other quarters too.

Like meat, wool is a form of animal protein. To insects like the clothes moth, the carpet beetle, the tapestry moth and the case-bearing clothes moth there is no more attractive diet than an appetising woollen jumper. Between them, the larvae of these insects do an immense amount of damage every year. They consume a vast amount of wool, and in doing so they cause an immeasurably greater amount of damage by ruining expensive made-up goods. A tiny hole chewed by a moth grub in a Parisian creation, for example, can cost the owner of the garment a small fortune. The clothes moth is the most destructive of all wool-loving insects.

The moth itself does not do any damage; its sole aim in life is to procreate and ensure survival of the species. The female moth lays her eggs in a suitable place, and within little more than a week the

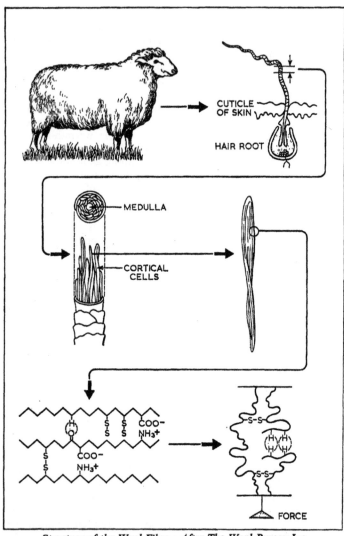

CUTICLE
OF SKIN

HAIR ROOT

MEDULLA

CORTICAL
CELLS

FORCE

Structure of the Wool Fibre – After The Wool Bureau Inc.

eggs will hatch into little white grubs. It is at this stage that the moth does all the damage. The grubs settle down and begin to chew their way through the wool in which their mother has so thoughtfully deposited them. The grubs may remain at this stage for as long as three years; after this they are ready to turn themselves into pupae and then moths, and are ready to start the whole life cycle all over again.

Anti-Insect Processes

The powerful synthetic insecticides developed in modern times are being used effectively against moths and beetles that attack wool. Hundreds of chemicals have been examined for this purpose, but only a comparatively few have been used industrially over significant periods of time. All are organo-chlorine insecticides stemming from the researches that led to the discovery of DDT. They include the mothproofing agents Dieldrin, Eulan U33, Mitin FF and Eulan WA new.

The insecticide Dieldrin is one of the most effective agents against moths and beetles. It is relatively simple and inexpensive to use, being applied as an aqueous emulsion that may be added to the liquors used in normal textile processing operations such as dyeing. Applied in this way Dieldrin is completely sorbed into the wool fibres and is removed only very slowly by washing. It resists removal by drycleaning agents such as white spirit and perchlorethylene.

Some of the organo-chlorine insecticides are equipped with chemical groupings that enable them to anchor themselves chemically to the wool fibre in the same way as dyestuffs. They are resistant to washing, drycleaning, light, rubbing, pressing etc. They are odourless and colourless, and produce no undesirable changes in the properties of the wool.

The use of these insecticides has come up against the difficulties associated with their use in other fields. Many are toxic materials if absorbed in significant amounts. Although the possibility of acute poisoning from wool treated with such materials is negligible under normal circumstances, there is the possibility of slow   sorbtion through the skin with prolonged contact producing chronic effects.

Storage

Wool is attacked by mildews which may damage and ultimately destroy the fibres. Stored in a badly ventilated warehouse, a damp and slightly alkaline wool offers ideal conditions for mildew attack. Chemicals such as sodium pentachlorophenate, mixtures of certain copper and chromium compounds and proprietary substances such as Shirlan NA can be used as mould preventatives, but wool is best stored dry in a well-ventilated place even when these materials have been applied to it.

Shrinkage

The scaly surface and the elasticity of the wool fibre play an important role in the shrinkage of wool fabrics and garments. Shrinkage can take place in two ways, by relaxation or by felting.

RELAXATION SHRINKAGE is caused by the tensions introduced during spinning, weaving and finishing. Wool is an elastic fibre, and will stretch when pulled. As fibres are being spun into yarns, yarns knitted or woven into fabrics and fabrics subjected to finishing treatments, the wool fibres are stretched. In the finished garment, the fibres sometimes retain some of this stretch; they have a tendency to try and return to their original unstretched state.

So long as the garment is kept dry, the fibres will be unable to recover their normal sizes and shapes. But once the wool becomes wet, the fibres are softened and lubricated and are able to relax into more natural positions.

A garment made up from wool in this sort of 'touchy' condition will therefore shrink when it is washed, no matter how carefully it is treated. A shower of rain will often provide sufficient moisture to loosen the fibres and cause relaxation shrinkage.

FELTING SHRINKAGE takes place in a different way. Unlike relaxation shrinkage, felting is unique to wool and other animal fibres.

Felting takes place when a wool fabric is subjected to mechanical action when it is wet. The fabric shrinks, but it also undergoes characteristic changes in its structure. The fabric becomes thick and the fibres are matted into closely-packed masses. The outline and character of the yarn pattern in the fibre become indistinct, and the fabric loses much of its elasticity. The surface of the fabric is covered by fibres, and its appearance is altered.

Although much research has been carried out on the felting of wool, the exact cause is not known. It seems almost certain that the characteristic scaly surface structure of wool and other animal fibres is the most important factor in felting.

The scales on a wool fibre are arranged with their edges pointing in one direction along the fibre. This means that a wool fibre is rougher in the tip-to-root direction than in the root-to-tip direction. A single fibre would therefore be able to move more easily through a mass of fibres in a root-first direction than in a tip-first direction. (There is some analogy with the movement of a head of wild barley placed inside a jacket sleeve; if the barley is inserted so that the hairs are pointing downwards, the barley head will move upwards through the sleeve. The bristly hairs will allow the head to move only one way; it is only movements influencing the barley in this direction that have any effect.)

Felting appears to take place most readily in a wool fabric when it is subjected to any treatment which causes the fabric to be repeatedly compressed, and then allowed to relax when it is wet. This treatment tends to bend the fibres into loops inside the fabric, and it has been suggested that felting occurs when 'travelling fibres' penetrate these loops to form knots and entanglements.

It has been shown that wool fibres will felt more readily when the tips of the fibres are softened or the roots hardened. If the tips are softened, the fibres are able to form loops more easily, and if the roots are hardened the travelling fibres can penetrate more readily into the loops.

The tumbling or stirring actions of washing machines, the prolonged use of the 'posser' (familiar in the north of England) and particularly any form of rubbing will all cause felting in wool. Some types of indoor dryer, which tumble the damp fabrics during drying, may also cause felting. Some wool fabrics felt more easily than others – a fluffy hand-knit sweater, for example, will felt more easily than a tightly-woven worsted flannel skirting.

Felting and Milling

This felting characteristic of wool fibres is used to advantage in the production of 'felts'. These are simply masses of fibres held together entirely by their natural holding-power without being spun or woven. Some types of woollen fabric are woven and then felted deliberately by the process known as *milling* to give them a fluffy, matted texture. Blankets, for example, are often felted to the stage at which the weave is indistinct before they are given a final raised finish. Deliberate felting is used in making fabrics such as beavers, meltons, doeskins and velours.

Washing

Compared with many fibres, wool is easy to wash; dirt and grease can be removed from it without difficulty. It is essential, however, that wool garments should be washed with care.

The temperature of the water used for washing wool has little or no effect on the shrinkage of wool garments in domestic washing. Water temperature does have an important effect on the fastness of wool dyes; hot water increases the risk of colours running and of white fabrics going yellow. The ideal temperature of water for washing wool is about 38°C (100°F), i.e. lukewarm. When a garment is soiled with heavy grease, rather hotter water may occasionally be used.

Washing Agents

Most of the washing products sold for domestic use are suitable for washing wool, and the type of product used will have little effect on shrinkage. Some products, however, will be more likely than others to cause colour bleeding and the yellowing of whites. In this respect, liquid synthetic detergents are generally the safest washing products, followed by soap flakes and detergent powders, and powders based on soap. There is no very great difference, however, between the 'safety factors' of these products with respect to their use in washing wool.

The manner in which the washing agent is used is very much more important than the agent itself. Solid agents must be completely dissolved before the wool fabric is put into the water. Undissolved particles of soap or other detergent may adhere to the garment and cause localized colour loss.

The actual concentration of washing agent used under normal conditions does not have a serious effect on the bleeding of wool dyes (excessive concentrations may cause damage). The correct amount to use is normally indicated on the package.

Hypochlorite bleaches must not be used on wool.

Minimum Handling

The washing procedure for wool articles should always be such as to involve a minimum of mechanical action. Mechanical action is the main cause of shrinkage; prolonged agitation, tumbling or stirring will cause felting. Above all, wool should never be rubbed.

The best way of washing wool garments is to leave them to soak for a minute or two in the washing solution. They may be squeezed gently by hand in the solution, and then left again undisturbed for a few minutes. This cycle of soaking and squeezing may be repeated several times.

The mechanical action necessary to remove dirt from blankets, without causing felting, can be applied by a wringer. After soaking for a few minutes in the washing solution, the blanket is passed through the wringer. This cycle of soaking and wringing may be repeated several times, the liquid is squeezed from the blanket by the wringer carrying away the dirt without felting the wool.

This soaking and wringing technique can be used effectively with other wool fabrics. Socks, sweaters and knitted garments, for example, can be washed safely in this way. (Knitted garments should be supported carefully when being handled, to ensure that no part of the garment is dragged down under it own weight.)

Close-fitting garments such as cardigans or sweaters are not affected seriously by any slight changes in shape caused during washing. Garments which hang to a definite length, however, such as jersey dresses, need extra care during washing. Jersey dresses should be dry-cleaned if possible. If they are washed, they must not be allowed to stretch by hanging when wet, or subjected to any twisting or rough handling.

Machine Washing

The stirring and tumbling actions of washing machines are not ideal for woól. Delicate woól garments, such as jersey dresses, fashion sweaters and loosely-knit fabrics should be washed by hand if possible. More robust fabrics, such as blankets, flannel shirts and shorts, socks or underwear may be washed in washing machines so long as the washing time is kept to a minimum. Even so, the vigorous mechanical action may cause felting after repeated washing.

A useful method of washing wool garments in a washing machine is to follow a soaking-agitation cycle which may be repeated several times. The load of wool garments is chosen so that there is no danger from colour-bleeding. After soaking for a minute or two in the washing solution, the garments are agitated for five seconds. The paddle is then switched off and the garments are allowed to soak again. After going through one or two soaking-agitation cycles in this way, the garments are passed through the wringer.

Wool garments should not be allowed to become very dirty before being washed. Wool can withstand frequent gentle washing without taking any harm. If a wool garment *does* become heavily soiled, it should not under any circumstances be subjected to an increased period of agitation in the washing machine. When the normal gentle methods fail to dislodge the dirt, the garment should be left to soak in cool suds for several hours, or even overnight.

This treatment is perfectly safe on undyed garments, or on single-colour fast-dyed articles, particularly when a liquid detergent is used.

Single-colour articles in which the dye tends to run a little will not, as a rule, suffer any serious loss of appearance when soaked in this way. There may be some lightening of shade.

Multicoloured wool fabrics should, however, be treated with especial care, as there is always the danger that dyes will bleed and affect another part of the garment which is white or of a different colour.

Rinsing and Drying

Wool garments should be rinsed in several changes of water after washing. They should be handled as gently as they are during the washing itself, and may be passed through the wringer between rinses, or squeezed carefully by hand. Knitted garments should never be twisted, or the fabric will be distorted.

After the final rinse and squeeze, the garment may be folded in a towel and passed through the wringer, pressed by hand or by being knelt on. This will remove much of the water that remains in the fabric.

Robust wool garments, such as socks, flannel shorts and skirts or blankets can be dried on a line in the usual way. More delicate fabrics and fashioned garments – such as jersey dresses, sweaters and cardigans, in which shape is important – should not be hung up to dry. These garments should be dried flat, away from intense heat. Sometimes, elaborate precautions are taken to return the garment to its original dimensions; the size and shape may be marked out on a piece of paper, for example, and the garment adjusted to this out-line before drying. Special 'shapes' made from wire or plastic may be used as formers on which wool garments can be dried to their original dimensions.

Drying flat has an additional advantage in that it keeps colour-running to a minimum during drying.

When drying is done out-of-doors on a line, white woollies should be hung in the shade to avoid yellowing which may be caused by direct sunshine.

Shrink-resist Processes

Shrinkage and felting have been studied by textile scientists for many years in an effort to improve the washing characteristics of wool. Many processes have been developed and some have been brought into practical use by the wool trade.

London Shrink Treatment

Relaxation shrinkage can be prevented by ensuring that the tensions in the fabric are eased before a wool garment is sold. The well-known London shrink treatment consists in wetting the cloth and allowing it to remain in a slack condition for a few hours. The fibres are then able to relax and the cloth shrinks before it reaches the customer. Fabrics treated in this way are described as 'London-shrunk'.

Although pre-shrunk fabrics of this sort will have lost much of their tendency to relaxation shrinkage, they will still felt and shrink if treated harshly during washing. To avoid felting, it is necessary to make some fundamental changes in the fibre-properties which cause it.

Anti-Felting Treatments

The felting of wool fibres which occurs during agitation in water is caused primarily by unidirectional movement of the fibres. This results in large measure from the 'stacked flowerpot' configuration of the scales on the fibre surface, causing greater resistance to fibre movement in one direction than in the other.

Anti-felting treatments developed for wool have evolved from attempts to modify the fibre surface in such a way as to minimise this unidirectional movement.

(1) Chemical Modification of Fibre Surface

In the early days of anti-felting research, processes were developed which aimed at the removal or smoothing-down of projecting edges of the surface scales on the fibres. Chemical teatments were often harsh, causing significant degradation of the fibre and damaging the cortex as well as the surface scales. Later research showed, however, that reducing the frictional differences by eroding the protruding scales was not essential in preventing unidirectional movement of the

119

fibres. Anti-felting properties could be realised by increasing the general level of fibre friction, so making fibre movement more difficult. Chemical anti-felting treatments were developed which involved comparatively mild attack on the fibre structure.

Many chemical anti-felting treatments have been in commercial operation for half a century and more, and are now in widespread use throughout the world.

WET CHLORINATION, one of the earliest processes, has undergone many modifications and developments since it was introduced. Sodium hypochlorite is the active agent, serving as a source of chlorine which attacks and modifies the surface of the wool fibre.

The process must be used with care or chemical degradation can be rapid and uneven, causing unsatisfactory shrink resistance, unnecessary damage to the fibre and difficulties in dyeing.

DRY CHLORINATION is carried out by treatment of the wool with chlorine gas. Wool subjected to this treatment should not contain more than some 8 per cent of water. It is necessary, therefore, to dry the wool which contains about 16 per cent of water in its air-dry state. This adds to the operating costs, which are generally higher than those of the wet chlorination process. On the other hand, dry chlorination can result in very even treatment of the wool.

Many other processes are now established in which a variety of chemicals are used to attack the fibre surface and produce anti-felting effects. Each has its own particular merits, the choice of process depending on the type of fibre, the manner in which it is to be processed and the intended use.

(2) *Addition of Polymeric Materials to the Fibre*

Anti-felting processes involving a chemical attack on the fibre must inevitably bring about changes in the characteristics of the wool. By removing part of the fibre surface they may cause significant loss of weight, and they can affect the strength and physical properties of the wool.

Since World War II much attention has been directed towards the development of anti-felting, shrink-resist processes in which material is added to the wool, notably by the formation or deposition of synthetic polymers in and on the fibre.

During the 1960s a wide range of inexpensive water-soluble or emulsifiable polymers became available, and many were examined as potential anti-felting agents for wool, with varying degrees of success. A particularly effective technique combined pretreatment with chlorine followed by the application of polymer from emulsion. This process could be used for the continuous treatment of wool tops – a more difficult problem than the treatment of wool fabric.

From this work have come modern processes culminating in the commercial production of fully shrink-proof wool, including the chlorine-Hercosett treatment used effectively on tops, loose wool and knitted garments.

Superwash Wool ·

The development of a range of processes now permits the textile manufacturer to produce any desired degree of shrink resistance in finished wool products, including completely shrink-proof materials. The degree of treatment provided depends upon many factors, including cost, type of wool, fabric construction, the demands of fashion, and the launderability requirements of the finished garment. Inevitably, this situation has created difficulties in the defining of shrink-resistance. What is shrink-resistant under one set of circumstances is not necessarily shrink resistant under other conditions. Moreover, the modern consumer, brought up in a world of synthetic fibres, has become accustomed to regard machine-washability as almost an essential requirement of everyday clothing and materials. A garment marked as shrink-resistant is expected to withstand machine washing.

To rationalise this situation and remove all risk of felting occurring during laundering, the International Wool Secretariat has prescribed high standards of shrink resistance and introduced the concept of Superwash wool. Goods treated for shrink resistance must conform to stringent specifications established and controlled by International Wool Secretariat before qualifying for labelling as Superwash. This indicates that the garment is fully machine-washable.

The Superwash appellation does not refer to any specific anti-shrink treatment; it signifies a standard of performance which may be achieved by treating wool with one of a number of commercial processes.

The Setting of Wool

It has been known since the earliest times that wool and other animal fibres can be formed into a desired shape and then persuaded to retain that shape for a period of time, perhaps short or perhaps long. Before the days of recorded history, for example, women were winding their hair around curlers and then allowing it to dry, when it remained in its curled shape. The effect was only temporary; when the hair became wet again, the curls fell out and the hair returned to its original shape.

More permanent forms of setting, as this process became known, have been used in the textile industry for a very long time, usually involving the simultaneous use of water and heat. Three well-known processes of this sort are crabbing, blowing and potting.

In CRABBING, cloth is wound under tension on a roller and steeped in boiling water, Stresses and strains introduced during weaving are dissipated and the fabric is set in a more stable form.

BLOWING is a related process, usually applied later in the finishing routine. Again, fabric is wound on a roller, usually interleaved with a smooth-surfaced wrapper, and subjected to the action of steam.

In POTTING or ROLL BOILING, fabric is wound under tension on a perforated roller, covered with a cotton or canvas wrapper and placed vertically in water which is slowly brought to 60 degrees C. It is kept at this temperature for times varying from 3 hours to 3 days, producing a cloth with a soft handle and smooth glossy surface.

The set introduced into wool fabrics by simultaneous use of water and heat in this way is more permanent than that produced by setting in cold water. The set shape will usually be retained if the fabric is wetted. But the set will be released if the wet fabric is heated to a temperature higher than that used in the setting process.

It is apparent, therefore, that there are different degrees of setting of wool, and in practice, set is graded empirically into three degrees; cohesive, temporary and permanent set.

COHESIVE SET is set that disappears when a fibre of fabric is released in cold water.

TEMPORARY SET is set that persists in cold water but disappears on release in boiling water.

PERMANENT SET is set that remains in the fibre or fabric even after release in boiling water.

The introduction of synthetic thermoplastic fibres led to the development of fabrics and garments capable of being set by the action of heat into desired configurations that are retained during subsequent wear and laundering. Garments can be set with permanent pleats and creases, and with permanent smoothness that resists the formation of wrinkles. These characteristics, which make for easy-care garments, were accepted gratefully by the consumer, who found them attractive and desirable. The establishment of the minimum-care concept stimulated interest in setting processes that could provide similar characteristics in traditional fibres, including wool.

The features of easy-care garments that make them what they are include the ability to withstand normal wear, dry cleaning and laundering. The aim has been, therefore, to achieve a degree of set as high as possible, with the ultimate target the production of permanent set.

Industrial Setting Processes

Research into the setting of wool fibres, extending over many years, showed that the rapid achievement of a high degree of set required the simultaneous action of a reducing agent, heat and water. These three influences could disrupt the bonds holding wool molecules together, permitting them to take up new positions determined by the shaping forces, and to re-establish bonds between the molecules that would hold them in their new positions.

Many industrial processes for setting wool fabrics and garments have been developed using the principle of simultaneous application of reducing agent, heat and water. Permanent pleats and creases set in wool garments are a match for those set in synthetic fibre garments. The processes used in setting wool differ from one another in the choice of reducing agent and the conditions under which it is applied; the agent may be introduced during finishing, or sprayed onto the finished garment prior to the final pressing or steaming operation.

In any textile manufacturing procedure, anything that adds to the production schedules is apt to cause complications and increase costs. Every effort has been made, therefore, to develop wool setting processes that will cause as little disruption as possible to normal manufacturing procedures.

SI-RO-SET Process

The Si-Ro-Set process, developed in Australia established a chemical spray-on technique for setting wool garments that has become the basis of subsequent processes. Permanent creases may be set in trousers, skirts and other garments by spraying a solution of a reducing agent – such as ammonium thioglycollate, mono-ethanolamine sulphite or sodium bisulphite – onto the garment prior to the final pressing operation, Processes of this type are now in widespread use; they are generally simple, cheap and effective.

Presensitizing Technique

The spray-on techniques described above involve the garment manufacturer in an extra manufacturing stage. With the object of avoiding this, setting processes have been developed in which the reducing agent – notably sodium bisulphite – is added to the fabric during finishing. Setting takes place during the garment manufacturer's final pressing operation, and he is no longer involved in the application of reducing agent to the garment.

Dry Processes

The need to add water to a finished garment prior to setting can also create difficulties for the clothing manufacturer, and many attempts have been made to devise setting processes in which he is no longer required to wet the garment. Techniques have been developed in which reducing agents are applied to the wool fabric during finishing, together with a humectant – a chemical that attracts water. During the final steam pressing of the finished garment, the humectant causes condensation of water in the cloth, thus fulfilling the three fundamental requirements for setting – simultaneous action of reducing agent, water and heat.

Flat Setting

Any technique adopted in the establishment of an 'easy-care' concept is required to contribute two vital characteristics to the garment. On the one hand, it must ensure dimensional stability; on the other hand it must provide for the retention of appearance during wear and laundering.

In their normal state, wool fabrics cannot meet these requirements. Firstly, the felting power of wool may lead to shrinkage and a resulting change in the shape and dimensions of fabrics and garments. Secondly, some wool fabrics may distort and become wrinkled during laundering.

To achieve easy-care characteristics in a wool garment, therefore, it is necessary to apply techniques that will overcome dimensional instability and loss of appearance during laundering.

Effective modern shrink-proofing processes are able to overcome the first obstacle. The second obstacle – how to ensure retention of appearance – may be overcome by applying established setting techniques to wool fabrics to stabilise them in flat form. Development of these techniques has centred around the application of water and reducing agent to a fabric and then subjecting it to steam treatment, for example on the blower. Sodium bisulphite is a reducing agent commonly used.

Modifications and variations of these techniques are now widely used. Combination of shrinkproofing with flat setting provides wool garments and fabrics which retain admirable dimensional stability and appearance under normal conditions of domestic use.

Permanent Press

The processes outlined above may be used effectively for garments that are required to withstand soaking in water or mild washing followed by drip drying. The garment remains in its desired shape for most of the time during laundering, and the set is sufficient to hold it in that shape.

If, however, a garment is to be subjected to severe machine washing and tumble drying, it will be for much of the time in a distorted shape. Changes in the fibres may take place which re-set them in their distorted shape during laundering. To prevent this happening and to achieve a really effectivce permanent set, a third requirement must be met in addition to shrink-proofing and setting. The wool must be stabilised in its desired shape.

This can be done in various ways. For example, the garment may be hung in a steam oven and subjected to superheated steam. Resin may be added to the fabric to encourage retention of the desired shape.

By making use of techniques now available, wool garments may be produced with outstanding easy-care characteristics that compare favourably with those associated with any synthetic fibre.

TECHNICAL NOTE

The keratin of wool is a protein of empirical formula $C_{72}H_{112}N_{18}O_{12}S$ (approx.). The exact composition depends upon the position of the material in the fibre and the treatment to which it has been subjected. Keratin is affected, for example, by the action of sunlight, and by the quality of the food which the sheep has eaten.

Keratin is hydrolysed by boiling with hydrochloric acid, yielding a mixture of some 17 α-amino acids of general formula $NH_2.CHR. COOH$. These acids are linked through their amino and carboxyl groups into a polypeptide chain as follows:

$$\begin{array}{ccccc} & \overset{R_1}{CH} & & & \overset{R_3}{CH} \\ & | & CO & NH & | \\ CO & NH & \underset{R_2}{CH} & CO & \end{array}$$

It is believed that there are about 600 amino acid residues in the main peptide chain of the keratin molecule. The molecular weight is probably about 68,000.

The strength of the wool fibre derives from these long molecules which lie alongside one another. They are held together by numerous strong hydrogen bonds between the carbonyl (C = O) and imino (NH) groups in adjacent molecules, and also by covalent disulphide and electrostatic salt linkages.

Disulphide linkages are formed by the amino acid cystine which has two amino groups and two carboxyl groups. When one of each of these groups is incorporated in adjacent polypeptide chains a disulphide bridge is formed: –

$$\begin{array}{ccc} CO & & CO \\ | & & | \\ CH . CH_2 . S . S . CH_2 . CH \\ | & & | \\ NH & & NH \end{array}$$

Salt linkages are formed by association between side chains on the polypeptide molecule, which contain free carboxyl and amino groups: –

$$\begin{array}{ccc} CO & & CO \\ | & & | \\ CH . CH_2 . CH_2COO^- \quad {}^+NH_3(CH_2)_4 . CH \\ | & & | \\ NH & & NH \end{array}$$

The unusual elasticity of the wool fibre may be explained by the coiled or folded state of keratin molecules in the relaxed fibre; this is alpha-keratin. When the fibre is stretched in steam, the molecules tend to uncoil or unfold; in this form, the material is described as beta-keratin.

126

THE PRINCIPLES OF SETTING

Supercontraction

During the 1930s a systematic scientific investigation of wool setting was carried out. W. T. Astbury and H. J. Woods studied the behaviour of wool fibres exposed to water and steam for varying periods of time whilst held in an extended state. It was found that fibres held at 40% extension and steamed for less than about 15 minutes would contract when released in steam to a length shorter than their original length. This phenomenon became known as supercontraction; it occurs to a maximum extent when the initial steaming is continued for only 2 minutes.

When fibres were extended and steamed for more than about 15 minutes, they retained a degree of permanent elongation when subsequently released in steam. The longer the initial steaming, the greater was the permanent increase in fibre length. This became known as permanent set.

Physical Mechanism

The action of steam on an extended wool fibre sets up two competing effects. Initially, some of the stabilizing cross-linkages and bonds in the fibre are broken. This permits the structure to relax and the fibre contracts into its foreshortened state. Disruption of these linkages takes place rapidly on steaming, but as steaming continues a cross-linkage rebuilding takes place, serving to stabilise the fibre in its extended state. On release after more than 15 minutes steaming, the fibre is no longer able to contract to its original length. Permanent setting has occurred.

In any practical setting treatment, the aim is to retain as much as possible of the deformation put into the fibre or fabric. This can be achieved in two ways: (1) by reducing the retraction forces that tend to pull the fibre back into its original state, and (2) by strengthening or increasing the number of new cross-linkages formed in the extended or distorted fibre.

One method of reducing the forces of retraction in an extended fibre is to hold it in water at an elevated temperature. The effect increases with increasing temperature and time; at 100 degrees C after 1 hour the retraction force is reduced by a factor of 30–40, so that only weak bonds are needed to set the fibre in its extended state. The degree of set given to a wool fibre during crabbing, for example, is much greater than that given to a woman's hair set in curlers at room temperature.

127

Chemical Mechanism

From his researches during the 1930s, J. B. Speakman concluded that the initial breakdown of wool on steaming involved hydrolysis of the disulphide cross-links in keratin molecules.

$$|\text{—CH}_2.\text{S—S}.\text{CH}_2\text{—}| + \text{H}_2\text{O} \rightarrow$$
$$|\text{—CH}_2.\text{SH} \quad \text{HOS}.\text{CH}_2\text{—}|$$

The second stage, i.e. rebuilding of cross-links, involved formation of new covalent bonds by reaction of some of the breakdown products with side chains in the wool molecule. This theory is still believed to be basically correct, the new cross-links contributing to the stability of permanently set wool, but the cross-links produced are now generally accepted as being different from those postulated in the 1930s.

Speakman suggested that the rebuilding of cross-links took place by reaction of sulphenic acid group formed as above with amino side chains to form –S–NH– cross links. Alternatively, and perhaps simultaneously, the sulphenic acid groups underwent further breakdown to form aldehyde groups which reacted in turn with amino side-chains. These reactions are as follows: –

$$|\text{—CH}_2.\text{SOH} + \text{H}_2\text{N—}| \rightarrow |\text{—CH}_2.\text{S}.\text{NH—}|$$

$$|\text{—CH}_2.\text{SOH} \rightarrow |\text{—CHO} + \text{H}_2\text{S}$$

$$|\text{—CHO} + \text{H}_2\text{N—}| \rightarrow |\text{—CH} = \text{N—}|$$

It is now generally believed that the disulphide bonds undergo a rearrangement stimulated by thiol groups introduced as a result of the action of reducing agents. These groups set up a thiol-disulphide interchange, breaking and reforming disulphide bonds which are thus rapidly rearranged into new positions corresponding to the extended state of the fibre. This interchange can occur slowly in boiling water, more rapidly in alkaline solutions and very rapidly in the presence of reducing agents.

This rearrangement removes the restoring force of the disulphide bonds originally present in the molecules and so prevents immediate contraction on release of a set fibre. But the new disulphide bonds are equally capable of rearrangement during release, and they do not make a significant contribution to the long-term stability of the set unless interchange is inhibited in some way after setting. This can be done by blocking or otherwise removing the thiol groups which catalyse the interchange.

It is now accepted that hydrogen bonds, which are present in great profusion in wool fibres, play a major role in the breakdown and rebuilding mechanisms that influence setting. Once the stabilising influence of the disulphide bond has been removed, massive hydrogen bond rearrangement occurs under the action of the boiling water.

This concept of setting suggests that fibres can be set at low temperatures by subjecting them to the combined acton of a reagent capable of breaking and rearranging disulphide bonds and one capable of breaking hydrogen bonds. This is the basis of low-temperature setting processes using, for example, solutions of reducing agents containing urea – a powerful hydrogen bond breaker. The reducing agent takes care of the disulphide bonds and the urea disrupts the hydrogen bonds, enabling the wool molecules to remain in the new positions they have taken up. If they can now be locked in these new positions, the fibre will have acquired a permanent set.

Many suggestions have been made as to the sort of mechanisms which can result in permanent set in this way. New covalent links may be formed, for example, by monosulphide lanthionine linkages created in the following way: –

$$|-CH_2.S-S.CH_2-| \quad \xrightarrow{\text{alkali}} \quad |-CH_2.S.CH_2-|$$

Another cross-link – lysinoalanine – is known to occur spontaneously in set wool as follows: –

$$-CH_2.S-S.CH_2-| \rightarrow |=CH_2 + S + {}^-SCH_2-|$$
$$|=CH_2 + H_2N(CH_2)_4-| \rightarrow$$
$$|-CH_2-NH-(CH_2)_4-|$$

Side chains other than lysine can also become involved in cross-links of this sort.

As these new cross-links cannot rearrange in the same way as disulphide bonds, they contribute stability to the set. On the other hand, if they should be formed before the necessary disulphide bond and hydrogen bond rearrangement has occurred, they can diminish the degree of set that takes place.

General Mechanism of Setting

A general mechanism of setting may be outlined which incorporates most of the accepted theories that have emerged.

Permanent set requires changes akin to melting; one crystalline form of wool keratin (alpha-helices) is converted to another (beta-crystallites). This necessitates rearrangement of relatively strong hydrogen bonds which requires the presence of water (or another hydrogen-bond breaker) and, usually, heat.

There is a simultaneous rearrangement of disulphide bonds generated, for example, by alkalis or reducing agents. And whilst these rearrangements are occurring, several new types of cross-link arise spontaneously. These new cross-links do not rearrange in the manner of disulphide bonds and they contribute to the stability of the set provided they are introduced after adequate hydrogen bond and disulphide bond rearrangement have occurred.

The various reactions involved in setting can be accelerated by heating at elevated temperature with steam under pressure. And additional stability can be obtained by hindering the reverse thiol-disulphide interchange that occurs during release by removing thiol groups or by introducing non-exchangeable cross-links after setting.

Insects which eat Wool

MOTHS   *Tineola bisselliella* (Common Clothes Moth)
        Tinea pellionella (Case-bearing Clothes Moth)
        Tinea pallescentella (Large Pale Clothes Moth)
        Trichophage tapetzella (White Tip Clothes or Tapestry Moth)
        Endrosis lactella (White Shouldered House Moth)
        Borkhausenia pseudopretella (Brown House or False Clothes
          Moth)
        Tinea lapella
BEETLES   *Attagenus piceus* (Black Carpet Beetle)
        Attagenus pellio (Furrier's Beetle)

Anthrenus scrophulariae (Common or Buffalo Carpet Beetle)
Anthrenus flavipes (Furniture Carpet Beetle)
Anthrenus verbasci (Varied Carpet Beetle)
Anthrenocerus australis (Australian Carpet Beetle)

In every case, it is the larva which causes damage by eating wool. The keratin of wool is an indigestible form of protein. It is believed that the larva secretes a substance in the middle intestine which reduces the disulphide bond linking the polypeptide chains together. This increases the solubility of the wool and enables it to be attacked more readily by protein-digesting enzymes.

A larva is able to cut a fine wool fibre with a single bite, but may have to gnaw its way through a thicker one. It is significant that larvae prefer finer wools.

Life Cycle of Common Clothes Moth

The life cycle of any insect is determined very largely by the conditions of its environment. Under ordinary circumstances, the Common Clothes Moth will begin to lay eggs the day after emerging from the pupa. It will continue to do so for a week to a fortnight; the adult moth will die within a fortnight to a month.

The eggs hatch in a week to a month. The larvae may persist for between three months to three years before turning into pupae. After 10–44 days the moth emerges from the pupa.

Moth Proofing

(1) Wool itself can be modified chemically to make it unpalatable to the grub. Reduction of the disulphide links followed by alkylation introduces a methylene group between the sulphur atoms. The new link is not broken down so easily by the digestive processes of the larva.

$$\left|-CH_2 . S . S . CH_2-\right| \longrightarrow \left|-CH_2 . SH \quad HS . CH_2-\right|$$

$$\longrightarrow \left|-CH_2S . CH_2 . S . CH_2-\right|$$

(2) The following chemicals are typical of those used for moth-proofing wool.

    (*a*) Martius Yellow. 2:4 dinitro-α-naphthol

    (*b*) D.N.O.C. Dinitro-o-cresol

131

(c) Inorganic fluorides
   e.g. Eulan Extra—a double fluoride of aluminium and ammonia.
   Eulan W Extra—potassium acid fluoride.
(d) Organo-Chlorine Compounds.

DIELDRIN

EULAN U33

MITIN F F

MOHAIR

Although wool is by far the most important animal fibre, there are a number of hair fibres which are of considerable commercial value. These come mostly from animals of the goat and camel families.

The Angora goat, which originated in Turkey, has a coat of long, lustrous hair which provides the textile fibre known as mohair. Much of the world's production of mohair is in Turkey, South Africa and the U.S.A.

Until early in the nineteenth century, Turkey was almost the sole producer of mohair. As the manufacture of textiles expanded during the period of the Industrial Revolution, efforts were made to raise the Angora goat in other parts of the world. Mohair production established itself in South Africa and in certain parts of the U.S., notably Texas and California towards the end of the nineteenth century. The U.S. is now the chief producing country and also the biggest consumer of mohair.

The goats are usually clipped twice a year, providing about 1.8–2.3kg (4–5 lb) of mohair per animal at each clip.

The quality of the fibre varies, depending on its source and the conditions under which the goat has lived. Fleeces are graded into tight lock, flat lock and fluffy types.

Tight lock is characterized by its ringlets and is usually very fine. Flat lock is wavy and of medium quality. Fluffy or open fleece is the least valuable.

As in the case of wool, mohair contains the dead, dull fibres that are known as kemps.

Fine Structure and Appearance

Like wool, mohair is contaminated with natural grease, dirt and vegetable impurities. These may account for as much as one-third of the weight of the raw fibre. The clean, scoured fibre is usually white and silky.

The fibres vary in length, depending upon the age of the goat. At six months, an Angora kid will provide fibres 10–15cm (4–6 in) long; at twelve months, they will be 23–30cm (9–12 in) long.

The surface of a mohair fibre has some resemblance to that of wool. It is covered with epidermal scales which are anchored much more closely to the body of the fibre than are wool scales. There are only about half as many scales as there are on wool. Mohair has some 5–6 scales per 100 microns length of fibre, whereas fine wool will have about 11. The overlap is very slight, so that the fibre has a smooth handle; light is reflected from the surface to give mohair its characteristic lustre. Individual scales usually encircle the whole fibre in the finer fibres.

As in the wool fibre, the bulk of the mohair fibre consists of a cortical layer built up from spindle-shaped cells. About one fibre in every 100 has a well-marked central core or medulla.

Mohair has a circular cross-section. Seen under the microscope, the cut end of a mohair fibre is marked by small spots or circles. These are caused by strings of air bubbles which lie between the spindle cells of the cortex.

133

Tensile strength

Tenacity: 11.8–12.8 cN/tex (12–13 g/tex).

Elongation: 30 per cent

Elastic Properties

Elastic Recovery from –
50 per cent breaking load: 0.8
50 per cent breaking extension: 0.6
Work of rupture: 2.65 cN/tex (2.7 g/tex).
Initial modulus: 353 cN/tex (360 g/tex).

Specific Gravity. 1·32

Effect of Moisture

Mohair absorbs water readily, and under normal conditions will hold as much moisture as wool.

Regain 13 per cent.

Effect of Heat, Age, Sunlight, Chemicals, Organic Solvents, Insects, Micro-organisms

Similar to wool.

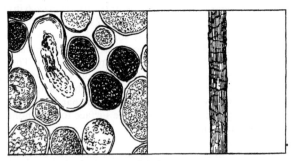

Mohair

MOHAIR IN USE

Mohair is characterized by its remarkable resistance to wear. Mohair fabrics are therefore used wherever durability is the first essential. Upholstery in public vehicles, car hoods, etc., are often made from mohair where resistance to wear and tear can be combined with attractiveness.

In combination with wool, mohair is often used for summer suitings. Mohair dyes well and has a beautiful natural lustre. It is made into all manner of dress materials, plushes and astrakhans.

Mohair is attacked by moths and should be protected in the same way as wool. It felts to a lesser extent than wool.

CAMEL HAIR

In north-west China and Mongolia, the Bactrian (two-humped) camel is an important animal. It serves as a means of transport in desert regions, and it also provides a supply of the camel hair which is used as a textile fibre.

Camel hair is shed by the animals in matted locks which are collected as they fall to the ground. Each animal yields some 2.27kg (5 lb) every year, and the total output of camel hair from China is more than 0.45 million kg (1 million lb) per year.

The camel has an outer coat of tough hairs that may reach 30cm (12 in) or more. Beneath this is a downy undercoat of fine soft hair 2.5–15cm (1–6 in) long. This camel hair undercoat is the really valuable part of the fleece. It is as soft and fine as merino wool. The downy wool is separated from the coarser hair by combing.

STRUCTURE AND PROPERTIES

Fine Structure and Appearance

Camel wool fibres are not so fine as cashmere; they are usually about 10–40μ wide. The surface of the fibre is covered with scales which cannot easily be seen under the microscope. The scales have diagonal edges.

The cortical layer of the camel wool fibre is marked by striations due to strings of coloured pigment granules that give the fibre its characteristic pale red-brown colour.

Some fibres have distinct medullae which are often fragmentary. Seen in cross-section, the fibres are circular or oval.

Tensile strength

Tenacity: 15.7 cN/tex (16g/tex).

Elongation. 39–40 per cent.

Elastic Properties

Elastic Recovery from –
50 per cent breaking load: 0.8
50 per cent breaking extension: 0.7
Work of rupture: 4.6 cN/tex (4.7 g/tex).
Initial modulus: 294 cN/tex (300 g/tex).

Specific Gravity.   1·32

Effect of Moisture.   Regain 13 per cent

Effect of Heat, Age, Sunlight, Chemicals, Organic Solvents, Insects, Micro-organisms

Similar to wool.

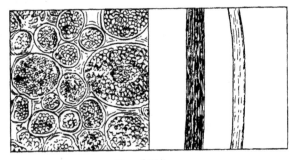

Camel Hair

CAMEL HAIR IN USE

Camel hair fabrics are warm and comfortable, and are used very largely for making overcoats, dressing gowns and knitted goods.

The coarse outer hair is made into ropes and industrial belting, tent-fabrics and blankets.

CASHMERE

In parts of China, northern India, Tibet, Iran and Afghanistan, the Tibetan cashmere goat is reared as a domestic animal. It has an outer coat of long coarse hair with an inner coat of down. This fine, soft down is the source of the fibre that is known all over the world as cashmere.

The downy cashmere fibre is combed from the goat's fleece during spring, and in the process is separated as much as possible from the coarser hair. Each animal will provide little more than 110g (4 oz) of cashmere fibre every year. The combined output from 30–40 goats provides enough fibre to make an overcoat.

STRUCTURE AND PROPERTIES

The downy fibres are usually 2.5—9.0cm (1–3½ in) long, and the coarser hairs 5.0–12.5cm (2–5 in) long.

Like other hair fibres, cashmere fibres are covered with epidermal scales. There are about 5–7 scales per 100μ on average. They have serrated edges and project from the fibre causing an irregular surface.

The cortical layer of the cashmere fibre consists of spindle-shaped cells with occasional long narrow spaces between the cells forming striations in the fibre. The fine cashmere fibre does not have any distinct medulla.

Cashmere wool is usually grey, buff-coloured, or white. The coloured fibres are full of tiny granules of pigment.

In cross-section, the fibre is circular or slightly oval; the pigment granules can be seen clearly in the cortical layer.

Cashmere wool fibres are extremely fine, averaging about 15μ in diameter, which is considerably finer than the best merino wool. The coarser beard hairs which are mixed with the true wool fibres are thicker, averaging about 60μ in diameter. The latter have well-marked medullae.

Chemical Properties

Cashmere is chemically similar to wool. It wets out with water much quicker than wool, and is more sensitive to the effect of chemicals, largely as a consequence of its greater fineness.

Cashmere is very easily damaged by alkalis such as washing soda. It will dissolve readily in solutions of caustic soda.

CASHMERE IN USE

The world output of cashmere wool is very small, and production is a costly process. Inevitably, the fibre is expensive.

Cashmere is so fine and soft, however, that it is regarded as one of the most desirable textile fibres of all. Cashmere fabrics are warm and comfortable, and have a beautiful drape. The fibre is used very largely for making high-quality clothes, shawls and hosiery, either alone or mixed with other fibres.

Garments made entirely from cashmere are properly labelled 'Pure Cashmere'. Sometimes, fabrics woven from fine botany wools are incorrectly described as 'cashmere'.

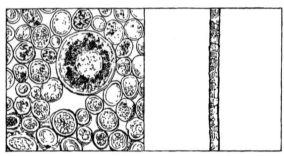

Cashmere

PERSIAN GOAT HAIR

The down from Persian goats is often marketed as Persian cashmere. It is, in fact, much coarser than true cashmere; the fibres are about 20μ in diameter. It is available in various shades of brown between a pale cream and a dark tan.

LLAMA

In the high mountain regions of the Andes, in Ecuador, Peru, Bolivia and north-west Argentina, the llama or South American camel is a beast of burden of immense economic importance. Like the Asian camel, it also provides a fleece which is a valuable textile fibre.

The fleece of the llama consists of a mixture of fine soft fibres and dull inelastic coarse hairs. The natural colours of the fleece are black, brown and white.

The fine fibres from the undercoat of the llama are not so fine as those of the Bactrian camel. Llama fibres are soft and strong, however, and often 30cm (12 in) or more long. They resemble camel hair in appearance; the fibre surface is scaly but the outlines of the scales are often difficult to detect. The medulla is pigmented.

Raw llama fleece contains only about 3 per cent of grease.

Llama hair is used locally, being made into carpets, rugs and hand-made clothing fabrics.

ALPACA

The alpaca is a close relative of the llama; it inhabits the same regions of South America. The animal has a soft fleece that may grow to a length of 60cm (24 in) if left uncut.

Alpaca fleece is strong, with an attractive glossy appearance when spun and woven into fabrics. It is generally finer than mohair but not so shiny.

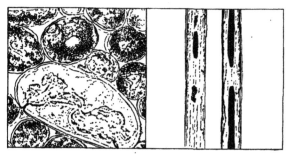

Alpaca

Alpaca has become an important article of commerce and the alpaca is now the most valuable of the fleece-bearing animals of the Andes regions. It provides much more and rather finer fibres than the llama.

The fleece colour is normally black, brown, fawn or white. The scales on the surface of the fibre are indistinct. The cortical layer is striated and there is rarely any medulla.

Alpaca is made into dress fabrics, linings, plushes and tropical suitings.

139

HUARIZO WOOL

The huarizo is a cross between the llama and the alpaca. Its fleece has a fine soft undercoat of lustrous fibes which are made into fabrics of first-rate quality. The outer hair covering is used for ropes and twine.

VICUNA

In Peru, there is a small species of llama called the vicuna, which provides a supremely fine fibre.

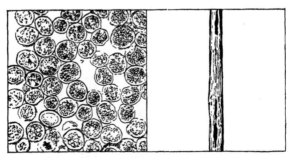

Vicuna

The vicuna runs wild in the Andes regions in Peru, and attempts at large-scale domestication have so far been unsuccessful. It is likely, however, that the vicuna will be domesticated in due course and vicuna fibres will become more plentiful. At present, the animal is shot for its fleece, which weighs only about 0.45kg (1 lb). The production of vicuna in Peru amounts to only a few thousand kilograms per year. The Peruvian government allows only a limited number of vicunas to be killed annually; there is little prospect of any increase in supplies until the animal has been domesticated successfully.

Vicuna fibre is generally regarded as being the finest and rarest wool-like fibre in the world. Its natural colour is white, fawn or brown, and garments made from it are usually left undyed. The fibres are about 5cm (2 in) long.

Vicuna fibres are only about half the diameter of fine wool fibres. The surface scales are regularly spaced and can be seen distinctly. The cortical layer is striated and there is rarely any medulla.

Vicuna wool is extremely expensive; it is made into superfine quality dressing gowns, coating materials and shawls.

GUANACO

The guanaco is a llama-like animal that roams wild over a large area of South America. It yields a wool that is similar to vicuna, but not so fine. It lies between vicuna and alpaca, with an average fibre diameter of about 20μ.

GENERAL PROPERTIES OF LLAMA-TYPE FIBRES

The llama, alpaca, huarizo, guanaco and vicuna animals are all related species, and their fibres are of a similar general type. They resemble the other hair fibres chemically and they behave accordingly towards processing liquors, bleaching and scouring agents, etc.

The llama-type fibres usually contain about 3 per cent of natural grease, with a total content of foreign matter amounting to perhaps 20 per cent.

The fibres are generally stronger than wool of comparable fineness. Their average diameters are as follows:

Llama	20–27μ
Alpaca	26–27μ
Huarizo	26μ
Vicuna	13μ

Llama, alpaca and huarizo can be compared with 55s to 60s wool, and vicuna with 120s or 130s wool (which are extremely rare qualities).

The surface scales of llama-type fibres are often difficult to detect under the microscope. The cortical layers are generally striated and filled with pigment granules in the case of coloured wools. There is usually a well-defined and unbroken medulla.

Llama-type fibres do not felt readily.

FUR FIBRES

The fur of animals such as the rabbit has long been used as textile fibre. There are two types of fur fibre; an outer coat of long, spiky fibres acts as a protective covering for an inner coat of soft, fine fibres which keeps the animal warm.

Angora rabbit-hair (often described erroneously as 'angora wool') has been in widespread use in European countries for a century or more.

The rabbits are clipped every three months; the fibres are 7.5cm (3 in) long. The outer 'guard hairs' are separated from the fine fur by blowing the fibres in a stream of air. Both hair and fine fur are used for making textiles, the former giving strength and beauty to the fabric and latter warmth and softness. The two types of fibre are mixed in such proportions as to provide the desired effect.

Other breeds of rabbits, including the wild rabbits that have become such a pest in Australia, are also used as a source of fur fibres. Much of the supply of rabbit fibre is used for making felts. The fibres are felted together without being spun or woven.

STRUCTURE AND PROPERTIES

The dimensions of rabbit fibres vary over a wide range. In general, the fine fibres are less than 20mm (¾ in), whereas the guard hairs may reach a length of several centimetres.

Seen in cross-section, the fine fur fibres are round, oval or rect-angular. The coarser guard hairs are often dumb-bell shaped, or in the form of a sharp-edged oval.

The scales on the surface of fine fur fibres are fairly uniform in shape. They often extend half-way round the fibre.

Scales on the guard hairs have serrated edges, and the edges often run slantwise across the fibre.

Both types of fibre have thick medullas, which contain many pockets of air.

Chemical Properties

The keratin of fur fibres is probably a mixture of several closely related proteins. The chemical behaviour of these fibres is generally similar to that of wool and other animal fibres.

Water is absorbed less readily by rabbit fibres than it is by wool. Hot water tends to soften or plasticize the fibres. Alkalis dissolve fur fibres.

FUR FIBRES IN USE

Rabbit hair and fur are used very largely for making felts and for knitted goods such as cardigans, gloves and berets. For knitted goods they are usually blended with wool before spinning.

Rabbit fibre fabrics have an attractive appearance and a soft luxurious handle. They wash like wool, and tend to felt very easily. This property is made use of in the manufacture of 'felts'.

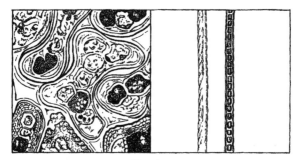

Rabbit Hair

SILK

Like wool, silk is an animal fibre. But instead of being grown in the form of hair, it is produced by insects as a handy material with which to build their webs, cocoons and climbing ropes.

Almost the entire commercial silk industry is based on one insect – the silkworm. In spite of its name, this is really a caterpillar; the silk is made by it when it wants to change into a chrysalis and then a moth. It spins the silk and wraps the fibre round itself in the form of a cocoon inside which it can settle down in comfort.

According to Chinese legend, silk culture dates back to the year 2640 B.C., when the Empress Si-Ling-Chi learned how to rear the caterpillars and unwind the cocoons that they made. The Empress devoted herself personally to rearing the worm, and it was largely through her encouragement that the silk industry became established in China.

For three thousand years China held a monopoly in the silk industry. Then sericulture – as silk production is called – spread to Japan via Korea. An ancient Japanese book – *Nihongi* – describes how in A.D. 300 a number of Koreans were sent from Japan to China to engage people experienced in the weaving and finishing of silk cloth. Four Chinese girls were brought back, and they instructed the court in the art of plain and figured weaving. The Japanese erected a temple to their honour in the province of Settsu.

Gradually, silk production spread westwards over Asia. Many tales are told of the ways in which the eggs of the silkworm and the seeds of the mulberry tree on which it fed were smuggled from one country to another. According to legend, they were carried to India by a princess who concealed them in her head-dress.

Between the Ganges and the Brahmaputra the Indian silk industry soon became established. From India, sericulture moved west again to Persia and the countries of the Mediterranean. Aristotle gives us our first description of the silkworm as 'a great worm which has horns and so differs from others. At its first metamorphosis it produces a caterpillar, then a bombylius and lastly a chrysalis – all these are changes taking place within six months. Then, from this animal, women separate and reel off the cocoons and afterwards spin them. It is said that this was first spun on the island of Cos by Pamphile, daughter of Plates.'

In this, Aristotle was probably incorrect, as raw silk was imported into Cos and woven some time earlier than this. The gauzy tissue that was made – *Coa vestis* – was notorious for its lack of covering power.

By the beginning of the Christian era, silk became one of the most coveted of all the treasures from the east; it was a royal cloth. The Emperor Justinian even monopolized the silk trade for himself, and looms were set up in the palace at Constantinople. Justinian learned of two monks who had lived in China and knew the art of sericulture. In A.D. 555 he persuaded them to return to China and bring back a supply of eggs from which a silk industry could start. The two monks made the journey and returned with silkworm eggs inside a hollow bamboo cane. From these eggs came the silkworms which established an industry in Europe that has lasted over a thousand years.

European Industry

By the eighth century, sericulture had been carried by the Moors to Spain and Sicily, and from there it spread to Italy and France. Florence, Milan, Genoa and Venice were all famous for their silks in mediaeval times. Silk weaving began at Tours in France in 1480; then in 1520, Francis I brought eggs from Italy and reared silkworms in the Rhône Valley.

During the reign of Edward III, the silk industry reached England. By the sixteenth century, silk was in widespread use among the wealthy nobility. Queen Elizabeth I wore silk stockings.

During the reign of James I (1603–25), silk workers from Italy were attracted to England in an effort to encourage the silk industry. In 1681, Charles II offered a grant of naturalization to weavers from the Continent, and in 1685 the revocation of the Edict of Nantes brought skilled French Huguenot weavers to England in their hundreds. During the twenty years 1670–90, some 75,000 immigrants were admitted into England. Many of them settled in the Spitalfields district of London; others went to Southampton, Bristol, Norwich, Canterbury and Sandwich.

By 1718 the first silk-throwing* mill in England was brought into operation at Derby. During the early part of the eighteenth century, other mills opened up at Congleton, Stockport and Macclesfield.

* 'Throwing' is a term that describes the twisting of long filaments of silk into a yarn. It is comparable with the 'spinning' of short fibres into yarn.

Since then, silk-throwing and weaving have been carried out continuously in Britain.

Silk in America

James I's efforts met with some success in 1619. Silk culture was encouraged by bounties; in their enthusiasm for sericulture the Virginians even expressed themselves in rhyme:

'Where worms and food do naturally abound,
A gallant Silken Trade must there be found.
Virginia excels the world in both –
Envie nor malice can gainesay this troth.'

In 1622, James I sent instructions to his colonists in America that 'they should apply themselves diligently and promptly to the breeding of silk worms, bestowing their labours rather in producing this rich commodity than to the growth of that pernicious and offensive weed, tobacco'. But James's efforts failed; 'despite the royal countenance the attempt was never attended by even partial success.'

Since then many efforts have been made to establish a silk industry in America. The high spot of the industry was in 1838, when a speculation fever gripped America. The South Sea Island mulberry, it was said, was especially suitable for feeding silkworms. Mulberry plantations sprang up like mushrooms overnight and America prepared to go in for silk growing in a big way. Young mulberry trees sold for many times their normal price, and speculators cashed in on the eagerness with which money was offered for anything to do with silk.

In 1839 the whole thing collapsed and the plantations were uprooted. By this time it was becoming apparent that silk cultivation was an economic proposition when plenty of cheap labour was available. Japan, China, India, Italy – these were the places where the necessary conditions prevailed. And it is in these countries that silk production has become important.

Japanese Supremacy

Before World War II, Japan was the world's leading silk producer – a position she established for herself in little more than half a century. After 1885, Japanese sericulture began on a big scale. During the fifteen years between 1892 and 1907 the output doubled; by 1930 it had become seven times that of 1892. The Japanese have devoted a great deal of study and intensive work to silk culture, and their success was largely due to the scientific approach they made to the many problems set by a unique industry.

During the present century, the silk-production industry has experienced severe setbacks. Viscose rayon took over much of the silk market during World War I, and had established itself when silk became available again after the war. During the 1920s and 1930s silk held its ground, particularly in the American hosiery trade. During World War II, however, nylon captured the ladies' stockings trade. Most of the silk producing countries were involved in the war; production in Japan, for example, was almost at a standstill.

Since the end of World War II, Japan, China, India, Bulgaria, Italy, France, Mexico, Turkey and other silk-producing countries have made great efforts to re-establish their position in the world's textile industry. But with new synthetic fibres joining nylon in the quality fabrics field it has proved an uphill task.

PRODUCTION AND PROCESSING

The silkworm is the caterpillar of a small off-white moth belonging to the species *Bombyx mori*. It lives on one thing only – the leaves of the mulberry tree. First essential for a silk industry, therefore, is an adequate supply of mulberry leaves – and the quantities needed are prodigious, for the silkworm spends its life doing little else but eat.

In Europe, silkworms are fed largely on the white-fruited mulberry, though other species are suitable. Much depends on the conditions under which the mulberries are grown, for this determines whether the leaves will be suitable for the worms.

Rearing of the silkworms starts as soon as leaves begin to appear on the mulberry trees. Eggs that have been laid by the moth and stored in a cool place during the winter are warmed up to encourage them to begin hatching. In large, scientifically-run farms, warming is done artificially; but where silk production is a part of peasant economy, as in many regions of China, the eggs are warmed by contact with the human body.

Hatching

After a few days the eggs hatch out to tiny caterpillars less than 3mm ($\frac{1}{8}$ in) long. 28g (1 oz) of eggs yields as many as 36,000 silkworms.

Every effort is made to get the eggs to hatch out in batches at the same time, as the economy of silk production depends largely on this; the worms will sleep and eat and spin at roughly the same time. Hatching is normally done by spreading the eggs over trays in the hatching shed. When the worms appear, perforated paper is placed over them and a supply of chopped mulberry leaves is spread on the paper. The worms climb through the holes and set to work on the leaves; dirt and egg-residues are left behind.

Moulting

During this stage the silkworms do nothing but eat, except for four periods of sleep lasting a day at a time, during which they shed their skins and grow new ones. Mulberry leaves are the sole diet. And they must be just right or the silkworm will refuse to eat them. They must be fresh and slightly wilted – but not faded.

After its fourth moult, the silkworm settles down to a final feed lasting approximately ten days, during which it eats twenty times its own weight of leaves. About thirty-five days have passed since it was hatched, and the worm is ten thousand times as heavy as when it was born. It is over 76mm (3 in) long and weighs about 7g (¼ oz). It has become a huge bloated greenish-white caterpillar filled with liquid silk, and is ready to start spinning. Rearing up on its hind legs, the silkworm weaves about looking for somewhere to settle down and build its cocoon. Bundles of straw are put on to the trays where the worms have been feeding, and those that are ready to spin climb ponderously up into the straw and begin to make their cocoons.

Spinning the Cocoon

The liquid silk is contained in two glands inside the silkworm. From these glands it flows in two channels to a common exit tube, called the spinneret, in the silkworm's head. As it emerges, the liquid silk hardens into very fine filaments and these are coated and stuck together by a gummy substance called sericin which comes from two other glands nearby.

The silk used by the worm, therefore, is really a twin filament held together as a single strand by the sericin cement. As the silk exudes, the silk worm moves its head backwards and forwards in a figure 8 movement. Gradually, it surrounds itself with a strongly built cocoon made from a continuous silk strand that may be up to 1.6km (1 mile) in length.

Spinning usually takes two or three days, during which time the silkworm has shrunk to a mere vestige of its original silk-bloated self. Inside its cocoon it begins to change into a chrysalis or pupa and then into a moth. On the silk-farm, the crysalis must be killed before this happens. For the moth escapes from the cocoon by secreting a fluid that dissolves away a section of the cocoon to make a hole through which it can crawl out. The continuous silk filament is thus broken up into thousands of short pieces which are useless for reeling. So within a few days of making its cocoon, the crysalis is killed by heating or stifling. It is baked in the sun or in an oven, or stifled in hot air or steam. The cocoon can then be kept indefinitely without damage until it is wanted for reeling.

Egg Production

From 28g (1 oz) of eggs, the rearer gets up to 63kg (140 lb) of cocoons. These yield some 5.5kg (12 lb) of raw silk. To produce it, the worms consume a ton of mulberry leaves. In many countries it is forbidden for silkworm rearers to provide themselves with eggs from their own moths. This is essential to control the many serious diseases to which the worm is prone. Egg production is thus an entirely separate branch of the industry and is carried out under rigorously controlled conditions.

The moths emerge from their cocoons as small greyish-white insects with rudimentary wings. They cannot fly; they have no mouths and cannot eat. The sole job in life of the silkworm moth is to mate and lay its batch of 350–400 eggs.

In order to check and control the health and vitality of the worms for spinning, each moth, after mating, is put into a linen bag 50mm (2 in) square. This has previously been cleaned and disinfected. After its eggs have been laid, the moth dies. Its body is examined microscopically, and if germs are present the bag and its contents are burned. In addition, some of the eggs – or 'seeds' as they are called by the rearers – are crushed and examined. It they are germ-free the eggs are passed for hatching.

Pests and Diseases

This careful control of silkworm eggs has been made necessary by the susceptibility of the silkworms to a number of epidemic diseases. For more than four thousand years the silkworm has been living an artificial life, and it has become as delicate as a hot-house plant.

A hereditary disease called *pébrine* is caused by a Protozoan parasite (*Nosema bombycis*). Infection of the silkworm causes black spots on the skin, and development of the insect is slowed.

Flacherie is a disease which may be due to digestive disorder or to an infective organism. It causes a swelling and blackening of the insect's body. It is not inherited.

Grasserie is a virus disease. The worms become yellow and bloated. They are filled with small crystals.

Muscardine is a fungus disease which kills the worm quickly. The silkworm's body is white and covered with spores. This is the most contagious of all silkworm diseases.

In hot countries, silkworms are attacked by a tachina fly (*Tricholyga sorbillaria*) which lays her eggs on the insect's body. As the eggs hatch, the maggots eat their way into the silkworm. They may emerge when the cocoon has been spun, eating through the silk and breaking the continuous filament into smaller lengths.

Pébrine

Worst of all the silkworm diseases is that known as pébrine. This first became serious at Cavaillon in France in 1850. It was extremely infectious and contagious and spread rapidly through the French silkworm industry. Within a year or two, France was having to import eggs from other countries to keep her silk industry going. But as the disease spread, production fell rapidly. In 1853 France produced 26 million kilos of cocoons; by 1865 this had fallen to 4 million. From France, pébrine spread to the silk industry of Italy, and within thirteen years of its occurrence the disease had resulted in a total loss of 120 million pounds to the silk trade in this area of Europe.

Gradually, pébrine spread to Asia and the Far East, until only Japan remained unaffected. But it proved less fatal in this part of the world; the life of the worms was more natural in these warmer countries, and they were more resistant to the disease.

To France, the pébrine epidemic was disastrous; but to the world it was to prove a blessing in a strange disguise. For it was through his investigation of pébrine that Louis Pasteur arrived at his germ theory of disease. Pasteur, in 1865, undertook a Government investigation of pébrine. He found that the disease was always accompanied by some sort of living corpuscles inside the body of the silkworm. These had been noticed before, but their direct connection with disease had not been suspected.

Pasteur found that the corpuscles – or germs as we now call them – were a characteristic of diseased worms. They were parasites living inside the body, and it was by transmission of such germs that the disease was spread.

Once the identification of these germs as the cause of the disease had been established, the way was open to control. Pasteur recommended that the stock from which eggs were hatched should be examined and strictly supervised. Any moth or eggs showing signs of germs should be destroyed, and every precaution must be observed to see that silkworm rearing was carried out cleanly and scientifically.

In this way, pébrine was brought gradually under control, and in many countries the silkworm industry today is run on hygienic and scientific lines that would do credit to a hospital.

Wild Silk

Although the domesticated silkworm *Bombyx mori* is the mainstay of the silk industry, there is a considerable trade in some countries in silk produced by silkworms living 'wild'. Most important of these wild silks is that which is known as Tussah ('Tussur', tussore).

Tussah Silk

Tussah silk is the product of several species of silkworm of the genus *Antheraea* – particularly *A. mylitta*, indigenous to India and *A. pernyi* which is native to China. These worms feed almost exclusively on the oak *Quercus serrata*.

Despite the fact that different species of worm produce Tussah silk, the cocoons are sufficiently alike for the silk to be regarded as a reasonably homogeneous material. The silk is affected more by the climatic conditions and the environment in which it has been produced than by the species of silkworm that produced it.

The tussah silkworm differs considerably in appearance and habits from *Bombyx mori*. It is usually larger, and may be 15cm (6 in) or more. It is a greener colour and is covered with tufts of gingerish hair.

The tussah silkworm lives an outdoor life, feeding on the leaves of dwarf oak trees. Crops of cocoons are produced twice a year, in the spring and autumn. The latter is the most important as a source of silk; the spring crop provides the worms that make the autumn cocoons.

Tussah silk production is an important peasant industry in northern China and Manchuria, and in parts of India. Manchurian cocoons are generally heavier than those from the Shantung region of China, and Manchurian silk is darker in colour. It has been estimated that 4 x $10^3 m^2$ (1 acre) of oak trees can support 60,000 cocoons; this is equivalent to about 360kg (800 lb) of raw silk.

151

The tussah silkworm leaves one end of its cocoon open, sealing the hole with a layer of sericin gum before settling down to its metamorphosis. When the moth wishes to emerge from the cocoon, it breaks through the sericin wall without damaging the continuous silk filament from which the cocoon has been made. Tussah silk cocoons are not necessarily treated, therefore, to kill the chrysalis before the silk is reeled.

Other Wild Silks

Although tussah silk is the most important wild silk in commercial use, there are other types of wild silk produced by caterpillars of different species in many parts of the world. In Japan, *Antheraea yama-mai* produces a silk that was at one time reserved for royal use. It feeds on oak leaves.

Attacus ricini provides a high-quality white silk; it is found in both the American and Asian continents. It feeds on the castor oil plant.

In Africa there is a silkworm of the Anaphe family which feeds on fig leaves. Groups of these caterpillars will build large nests inside which they make their individual cocoons. The nests and cocoons are made entirely from silk. This type of wild silk is collected in considerable quantities in Uganda and Nigeria, and used for making native fabrics.

In India, *Antheraea mylitta* and *Antheraea assama* are important silk-producing caterpillars. They make cocoons often more than 5cm (2 in) long. *A. mylitta* feeds on the bher tree, *Zizyphus jujuba*.

Reeling and Throwing

Vegetable fibres such as cotton, and hair fibres such as wool have one thing in common; they are produced in relatively short lengths. Cotton and wool fibres are usually a few centimetres long; even flax, which is one of the longest of the vegetable textile fibres, is only about 60cm (2 ft). In order to convert these short fibres into long threads or yarns, we have to align the fibres and then 'spin' them by twisting the fibres together. In this way, innumerable short fibres are made to grip one another to form a thread or yarn that is long enough to be used for weaving purposes.

Silk, however, is quite different from these other natural fibres; the silkworm makes its cocoon from a twin filament that is extruded from its spinneret in a continuous strand. This filament may be as much as 1.6km (1 mile) in length.

The production of a 'thread' or 'yarn' of silk suitable for weaving

ELEMENTARY PRINCIPLES OF RAW SILK REELING

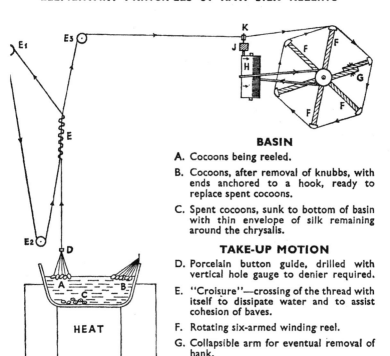

BASIN

A. Cocoons being reeled.

B. Cocoons, after removal of knubbs, with ends anchored to a hook, ready to replace spent cocoons.

C. Spent cocoons, sunk to bottom of basin with thin envelope of silk remaining around the chrysalis.

TAKE-UP MOTION

D. Porcelain button guide, drilled with vertical hole gauge to denier required.

E. "Croisure"—crossing of the thread with itself to dissipate water and to assist cohesion of baves.

F. Rotating six-armed winding reel.

G. Collapsible arm for eventual removal of hank.

TRAVERSE

H. Drum rotated by belt from winding reel.

J. End of sliding rod, pinned eccentrically to drum top, causing to and fro movement across the direction of the thread.

K. Guide eye, set on sliding rod, to spread the ends and so give width to the hank.

Courtesy: P. W. Gaddum

is therefore a process different from that which is used in the case of shorter fibres. All that is necessary, in principle, is to unwind the long continuous filaments from the cocoons and then twist a number of these together to form a thread of useful thickness.

The unwinding of the fine silk filaments from the cocoons is called reeling, and the process is carried out in a building called a filature. The cocoons are soaked in hot water to soften the sericin gum that is cementing the filament in place. A revolving brush is then used to find the end of the filament – a difficult job with something so fine that it is almost invisible.

When the end of the filament has been picked up, it is drawn through a guide along with the filaments from several other cocoons. The filaments may be given a slight twist to hold them together and are reeled steadily off the cocoons which are left floating in hot water to keep the gum softened.

Tussah cocoons are gummed more firmly than those of cultivated silk, and contain more calcium salts. They are usually soaked in sodium carbonate before reeling.

Reeling requires great skill, as the operator must produce a uniform thread by combining the silk filaments in suitable fashion. Each filament is narrower towards the beginning and the end than it is in the middle, and the reeler must join the filaments in such a way as to allow for this variation in width. The actual size of the threads produced is denoted by weighing a certain length in half decigrams. This is called the denier of the silk.

Silk is wound up by the reeler in the form of skeins. These are made up into bundles of about 2.7kg (6 lb), called 'books'. These are then packed into bales for shipment.

When the raw silk arrives at the manufacturing centre, it is in the form of continuous strands in which the individual filaments are cemented together by the sericin. Silk may, for some purposes, be woven without further preparation. Usually, however, two or three of these multi-filament strands are twisted together to form heavier threads; this process is called 'throwing' from the Anglo-Saxon word 'thrawan' meaning to whirl or spin.

Degumming

The natural gum, sericin, is normally left on the silk during reeling, throwing and weaving. It acts as a size which protects the fibres from mechanical injury. The gum is removed from the finished yarns or fabrics, usually by boiling with soap and water.

Silk fabrics woven with the sericin still on the yarn have a characteristic stiffness of handle; they are also dull in appearance. After degumming, the silk acquires its beautiful lustre.

As much as one-third of the weight of the fabric may be lost when the gum is removed in this way.

Raw silk with the gum still on the filaments is called 'hard silk'. Degummed silk is 'soft silk'.

Foulard fabric, georgette, chiffon and crêpe de chine are woven from hard silk which is afterwards degummed.

Types of Thrown Yarn

There are a number of different types of thrown yarn, which are described by the manufacturer as follows:

TRAM. This is a lightly-twisted thread formed by twisting two or three strands of silk together. Low twist tram will have only two or three twists to the inch (2.5cm); high-twist tram may have 12–20 twists to the inch (2.5cm).

Tram is moderately strong; it is soft and has a full handle. Tram yarns are used as weft in woven fabrics.

ORGANZINE. This is a very strong yarn made from high-quality silk. Two or more strands are each twisted and the compound thread then twisted in the opposite direction, from about 9–30 turns to the inch (2.5cm). Organzine is used mainly as warp in a woven fabric.

CRÊPE. These are yarns with a very high twist, as many as 30–70 to the inch (2.5cm). They are used for crêpe fabrics and chiffon, and for knitting to hosiery.

GRENADINE. This is a tightly-twisted yarn, in which two or three twisted strands are combined and twisted in the opposite direction. It is twisted more tightly than organzine.

Grenadine yarns have extra good weaving qualities and a delustred appearance. They are used for high-quality silk hose.

COMPENZINE. This yarn is made from two tightly twisted yarns and one untwisted yarn. When these are twisted together, about five turns to the inch (2.5cm), the untwisted yarn crinkles up, giving the 'knobbly' appearance characteristic of crêpe threads.

SEWING SILKS. These are tightly-twisted strong yarns. They are made by twisting two or three silk strands together and then combining several of the resulting threads by twisting in the opposite direction.

155

EMBROIDERY SILKS are often simple untwisted strands united by a slight twist.

Silk Wastes

Although the silkworm spins its cocoon from a continuous filament of silk, the throwster is fortunate if he can make use of half of the available silk in filament form. The rest of the silk is unsuitable for reeling, and is known as 'waste silk'.

This waste silk is much too valuable to throw away, and it is used for making the yarns we know as 'spun silk'.

Waste silk consists of the silk brushed from the outside of the cocoon during reeling, the unusable inner portions of the cocoon, broken filaments from damaged cocoons such as those from which the moth has been allowed to emerge, and waste material from the reeling and throwing generally. It is packed into bales, and arrives at the spinning mill as a mass of filaments of different lengths, contaminated with dirt and straw.

The waste silk is cleaned and degummed in one of two ways. In the *English process* of degumming, the gum is removed by boiling the silk in soapy water. This dissolves the sericin and leaves a clean, smooth filament. The *Continental process* uses a fermentation technique, and about 20 per cent of the gum remains on the silk. Waste silk degummed in this way is called 'chappe' or 'schappe' silk; it is used for making velvet.

After degumming, the silk filaments are subjected to processes similar to those used for wool and other short-staple fibres. The silk is opened and loosened in a machine that delivers it in the form of a gauze-like blanket or lap. The fibres are then combed and sorted into length-groups, and then drawn into rovings and spun by twisting so that the short fibres hold tightly together.

Spun silk yarns are used for making scarves, ties, velvets and pile fabrics, woven dress materials, knitted goods, lace, shirts and a variety of union fabrics.

Silk Counts

NETT SILK (Filament yarn). The fineness of a silk yarn is denoted by its 'denier'. This is the weight in grams of 9,000 metres of the yarn. The lower the denier, therefore, the finer is the silk.

SPUN SILK. Spun silk yarns are defined by counts in the same way as cotton yarns. The count is the number of hanks, each 840 yards (756m) long, that will weigh 1 lb (454g).

Dyeing

Silk is dyed very largely in the form of hanks or woven pieces. There is an immense range of dyestuffs available for use with silk; almost every class of dyestuff used for cotton or wool can be used for dyeing silk. In general, the dyestuffs are applied by techniques similar to those used for wool or cotton.

Spider Silk

The silken filaments spun by spiders are so fine that they can often be seen only with difficulty. The golden garden spider spins a filament only 0.00001 in (0.00025mm) in diameter.

Many attempts have been made to use spider silk as a textile fibre. More than 200 years ago, a Monsieur Bon of Languedoc in France collected enough silk from spider cocoons to spin into a yarn. He made silk stockings and gloves from the fine grey silk and exhibited these at the Academy of Science in Paris in 1710.

Great excitement was aroused in France and M. Bon was loaded with honours. René Réaumur, the famous physicist, was commissioned to examine the feasibility of setting up a spider silk industry. But in spite of his initial enthusiasm, Réaumur concluded that the difficulties would be too great. The spiders were temperamental and unco-operative; they became excited and resented the food they received so much that they ate each other instead. Such silk as they produced was delicate and extremely difficult to spin.

In 1864, Dr. Wilder, an American army surgeon stationed in South Carolina revived the idea of using spider silk for textiles. He found that instead of waiting for the spiders to spin their cocoons he could 'milk' the silk artificially from the spider. A pair of stockings made from this spider silk cost over 100 dollars. They represented the life's work of nearly 500 spiders, and the stockings were so sheer that they were of little practical value.

Since then, attempts have been made in other parts of the world to domesticate the spider and relieve it of its silk. A 2½ in (6.35cm) spider living in Madagascar has been used. 'Milked' of its silk by native girls five or six times a month, it yields about 3.2km (2 miles) of filament and then dies.

A fabric of spider silk, 18 yd (16m) long and 18 in (45cm) wide was shown at the Paris Exposition in 1900. It contained 100,000 yd (90km) of thread containing 24 strands – the work of more than 25,000 spiders.

Today, we have virtually given up the attempt to use spider silk in textiles. It still finds an important outlet for making 'crosswires' in optical instruments. The silk is uniform and strong, and withstands changes in humidity and temperature. Spider silks used in this way have often outlasted the life of the optical instrument, remaining unchanged after half a century or more.

STRUCTURE AND PROPERTIES

Fine Structure and Appearance

The raw silk strand from which a cocoon is built consists of two fine filaments cemented together by sericin gum. Seen under the microscope, raw silk has a rough and irregular surface, and it is marked by lumps, folds and cracks in the sericin layer. Often, the twin filaments of silk are separated for considerable distances, each with its own coating of sericin.

Seen in cross-section, the strand of cocoon silk is of irregular shape. It is roughly oval with average diam. of 0.178mm (0.007 in). The individual filaments (brins) can be distinguished inside the sericin coating. They are triangular in cross-section, with rounded angles. Usually, the filaments lie with one flat side of each facing the other.

The degummed filaments are smooth-surfaced and semi-transparent. The diameter fluctuates from place to place, averaging 0.0127mm (0.0005 in). The filaments become thinner towards the inside of the cocoon.

In the raw state, silk varies in colour from cream to yellow. Most of this colour lies in the sericin gum, and is lost when the filaments are degummed. The silky sheen develops after degumming.

Wild Silk

There is naturally much more variation in the physical properties of wild silk than there is in cultivated silk. Colour, for example, may be yellow, grey, brown or green.

Seen under the microscope, wild silk may be distinguished from cultivated silk by its irregular width. It is also marked by longitudinal striations and tends to have flattened areas on which are transverse markings. These flattened areas are caused by filaments pressing against one another in the cocoon before the material of the silk has hardened.

Treatment of the wild silk filament with chromic acid will disintegrate it into a bunch of finer filaments, fibrils or micelles about $1 \cdot 0\mu$ in diameter.

The same effect can be obtained by severe mechanical or chemical treatment of cultivated silk, in which the fibrils are more closely compacted.

Seen in cross-section, the twin filaments in wild silks are wedge-shaped, with the short bases facing one another. The cut section of the filament is dotted with markings corresponding to the striations running lengthwise through the filament. These mark the boundaries between the fibrils, which are less closely held together than in cultivated silk.

Tensile Strength

Silk is a strong fibre. It has a tenacity usually of 30.9–44.1 cN/tex (3.5–5.0 g/den). Wet strength is 75–85 per cent of the dry strength.

Elongation

Silk filaments have an elongation at break of 20–25 per cent under normal conditions. At 100 per cent R.H. the extension at break is 33 per cent.

Elastic Properties

The elastic recovery of silk after spinning is not so good as that of wool, but is superior to that of cotton or rayon. Once it has stretched by about 2 per cent of its original length, silk tends to remain permanently stretched. There is a slow elastic recovery or creep after extension, but the silk does not regain its original length.

Elastic recovery from	*Japanese*	*Tussah*
50 per cent breaking load:	0.56	0.40
50 per cent breaking extension:	0.38	0.41
Work of rupture (cN/tex):	5.98	7.46
Initial modulus (cN/tex):	735.8	490.5

Specific Gravity

Degummed silk is less dense than cotton, flax, rayon or wool. It has a specific gravity of 1·25. Silk fabrics are often weighted by allowing the filaments to absorb heavy metallic salts; this increases the density of the material and affects its draping properties.

Effects of Moisture

Like wool, silk absorbs moisture readily. It can take up a third of its weight of water without feeling wet to the touch. Silk has a regain of 11·0 per cent.

159

If there are salts or impurities in the water, silk tends to absorb them too. Hard water, therefore, will contaminate silk. Degummed silk will swell as it takes up moisture; the extent of the swelling depends upon the relative humidity of the atmosphere. At 100 per cent R. H., silk absorbs 35 per cent of its weight of water and increases in cross-sectional area by 46 per cent (from 65 per cent R.H.).

Effect of Heat

Silk will withstand higher temperatures than wool without decomposing. Heated at 140°C. it will remain unaffected for prolonged periods. It decomposes quickly at 175°C.

Silk burns, emitting a characteristic smell like that of burning hair or horn.

Effect of Age

Silk is attacked by atmospheric oxygen, and may suffer a gradual loss of strength if not carefully stored.

Effect of Sunlight

Sunlight tends to encourage the decomposition of silk by atmospheric oxygen.

Chemical Properties

The strands of raw silk as they are unwound from the cocoon consist of the two silk filaments mixed with sericin and other materials. About 75 per cent of the strand is silk and 23 per cent is sericin; the remaining material consists of fat and wax (1·5 per cent) and mineral salts (0·5 per cent).

As would be expected in a fibre of animal origin, silk is a protein. The filament itself is the protein fibroin; it is similar in composition to the sericin protein, despite the differences in physical behaviour between the two materials.

Silk fibroin differs fundamentally from the keratin from which animal hair fibres are made. The fibroin molecule contains only carbon, hydrogen, nitrogen and oxygen; there is no sulphur in it.

Silk does not dissolve in water, and it withstands the effects of boiling water better than wool. Prolonged boiling tends to cause a loss of strength.

Silk will readily absorb certain salts from their solutions in water; aluminium, iron and tin salts, for example, are used for 'weighting' silk fabrics in this way.

Silk dissolves in solutions of zinc chloride, calcium chloride, alkali thiocyanates and ammoniacal solutions of copper or nickel. Silk fibroin is attacked by oxidizing agents; bleaches such as hydrogen peroxide must be used with care. Hypochlorite bleaches must never be used, as they rapidly tender silk.

Effect of Acids

Like wool keratin, the fibroin of silk can be decomposed by strong acids into its constituent amino acids. In moderate concentration, acids cause a contraction in silk; this shrinkage is used to bring about crêpe effects in silk fabrics.

Dilute acids do not attack silk under mild conditions. Organic acids are used for producing the scroop of silk, which may be due to the surface-hardening of the filaments. Acids are readily absorbed into silk filaments, and are not easily removed.

Effect of Alkalis

Silk is less readily damaged by alkali than is wool. Tussah silk is particularly resistant. Weak alkalis such as soap, borax and ammonia cause little appreciable damage. More concentrated solutions of caustic alkalis will destroy the lustre and cause loss of strength.

Silk dissolves in solutions of concentrated caustic alkali.

Effect of Organic Solvents

Silk is insoluble in the dry-cleaning solvents in common use.

Electrical Properties

Silk is a poor conductor of electricity, and tends to acquire a static charge when it is handled. This causes difficulties during manufacture, particularly in a dry atmosphere, but is of value for insulating materials in the electrical trade.

Other Properties

Raw silk has a rough handle; it acquires its smooth 'silky' feel only when the gum has been removed.

The peculiar noise or 'scroop' made by silk when it is crushed is not an inherent property of the fibre. It is given to the silk by treatment with dilute acids, possibly through a surface-hardening effect. Scroop is not a sign of quality, as is commonly supposed.

161

SILK IN USE

For thousands of years, silk has reigned as the queen of fibres. It is expensive, tedious to produce and subject to all the hazards inevitable in an industry whose assembly line is a living thing. Yet until a few years ago, silk has been unchallenged in its position as the most desirable of all textile fibres.

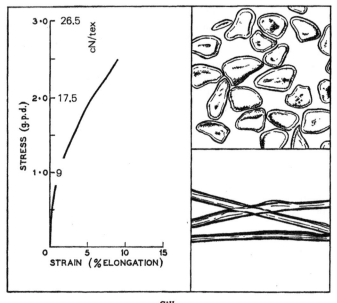

Silk

The combination of properties that has made this possible is all the more amazing in that the fibre is manufactured by the larva of an insignificant moth.

Silk combines a high strength and flexibility with good moisture absorption, softness and warmth, excellent wearability and a luxurious appearance.

Silk is so versatile that it is woven and knitted into a wide variety of fabrics; it provides all manner of materials from the sheerest chiffon to the richest of heavy pile velvets. Silk is cool and comfortable in underwear or summer clothes; it is hard-wearing and easy to clean in dresses and sturdy suitings. The smooth-surfaced filaments from which silk fabrics are made do not hold on to dirt.

Laundering

Despite its resistance to wear, silk is a delicate fabric. The filament of silk is fine and easily torn. It can be damaged by chemical action, and must be washed with care. Human perspiration can degrade the fibroin of silk, and silk garments should be washed regularly.

Silk should be laundered with soap flakes or a mild detergent, rinsed thoroughly in soft water, dried gently and ironed while damp. Wild silk fabrics should be ironed dry.

The plasticity of silk is made use of in finishes where pressure is applied to the fabric. The soft, warm filaments are pressed out of their normal shape, and fabrics can be given special glazed effects. If carelessly ironed, silk garments may acquire an undesired glaze in this way; the iron should be used at moderate heat, and a pressing cloth put between the iron and the silk. This will prevent distortion of the plastic silk filaments.

Crêped garments should be dried after gently stretching to their original shape. Ironing should be done on the wrong side of the fabric, with a blanket as support.

Weighted silks are often affected drastically by washing or dry cleaning. This may be caused by removal of the metallic salts from the fibre. The effect can be restored to some extent by dipping the fabric in a dilute solution of gum arabic.

Weighting

Silk is so costly that it is rarely made up into heavy fabrics of pure fibre. Many techniques have been devised for increasing the density of silk by artificial means; the weighting of silk with metallic salts such as tin chloride, for example, has long been in use, though not now to the same extent as before.

Degummed silk is steeped in a solution of tin chloride, and the silk filaments absorb some of the salt. After washing, the silk is steeped in a solution of sodium phosphate. This double steeping process is repeated several times, after which the silk is soaked in a bath of sodium silicate solution.

The quantity of metallic salt absorbed by the silk in this way is adjusted over a wide range. A moderately weighted silk will contain 25–50 per cent of salt.

A crêpe-de-Chine is normally weighted with 25–45 per cent of

163

salt; satin may contain 50 per cent and georgette 30 per cent. Heavily weighted tie fabrics will contain as much as 60 per cent of absorbed material.

In general, weighted silk fabrics are not as strong as those made from pure silk. They have a fuller handle which is often preferred for certain applications. Heavily weighted fabrics are more sensitive to the effects of light and air, and deteriorate rapidly under certain circumstances. Perspiration, for example, will attack weighted silk and make it rot; loaded silk garments should not be used as underwear.

TECHNICAL NOTE

Chemical Constitution

The fibroin of the silk filament is a protein similar in composition to the sericin gum surrounding it. Like wool and other proteins, silk fibroin is formed by the condensation of α-amino acids into polypeptide chains. It differs from the hair proteins in that it does not contain sulphur; the long chain molecules are not linked together by the disulphide bridge as they are in wool.

When silk fibroin is hydrolysed by treatment with strong acids, it yields a mixture of amino acids, including:

glycine	$NH_2 CH_2 COOH$	41·2 per cent	(grams of amino acid per 100 grams of protein)
alanine	$NH_2 CH CH_3 COOH$	33·0 per cent	,,
serine	$NH_2 CH COOH$ $\mid$ CH_2OH	16·0 per cent	,,
tyrosine	$NH_2 CH COOH$ $\mid$ $CH_2 C_6H_4OH$	11·4 per cent	,,

Molecular Structure

The molecular weight of fibroin is not known with any accuracy. It has been estimated as between 84,000 and 220,000.

X-ray analysis of silk fibroin has shown that there is a high degree of crystallinity in the silk filament. The polypeptide chains are able to pack together in such a way that they form regions of regularity typical of a crystal. The atoms of the molecules in these crystalline

164

regions are exerting natural forces of attraction which hold the molecular chains tight against one another.

In the silk filament, the polypeptide chains are fully extended, whereas in the wool fibre they are folded. Silk has, in this respect, a close resemblance to the stretched wool fibre.

The closely packed and aligned molecules of silk fibroin are associated into regions of high order or crystallinity whilst regions of disorder also occur where the fibroin is amorphous and the molecules are not orientated (see 'Cellulose'). Individual fibroin molecules may form part of several crystalline regions, and of the amorphous material in between.

The fully extended nature of the silk molecules and the high degree of orientation accounts for the low elasticity of silk. The molecules cannot unfold like wool molecules when a filament is subjected to a pull. There is only a small amount of distortion before slippage of the molecular chains takes place.

The close packing of the silk molecules and the high degree of crystallinity confer great strength on the silk filament. The natural forces of attraction between the molecules can operate with maximum effect.

C: NATURAL FIBRES OF
MINERAL ORIGIN

ASBESTOS

Asbestos is one of the strangest of all the naturally occurring fibres. It is a rock which has been subjected to unusual treatment during its formation. Instead of crystallizing in the normal way, it has done so in the form of fibres. These are all packed tightly together alongside each other, giving a grainy structure to the rock, resembling wood. Asbestos is non-inflammable and heat resisting.

Asbestos has been known and used as a textile since the earliest times. The lamps of the vestal virgins are supposed to have had wicks made from asbestos, and a lamp made for Minerva 'had a wick made of Carpasian linen, the only linen which is not consumed by fire'. This linen was made from asbestos mined in Cyprus.

The Chinese used asbestos to make false sleeves which could be cleaned by putting them in the fire. All the dirt was burnt off, leaving the asbestos clean. The Emperor Charlemagne is alleged to have owned a tablecloth of asbestos which he threw on to the fire after a meal, to the consternation and amazement of his guests. When Charlemagne was being threatened by Harun-al-Raschid and his hordes, he put his tablecloth to excellent use by performing his trick before the ambassadors of the fierce Emperor of the East. They were convinced that Charlemagne was a wizard, and recommended to Harun-al-Raschid that the impending war should be called off. So asbestos had an early influence on the history of the world.

In 1684, we hear of an asbestos handkerchief being demonstrated before the Royal Society of London. And after the discovery of asbestos in Russia between 1710 and 1720, a factory was established for the manufacture of asbestos textiles, socks and gloves. This industry continued for half a century.

Benjamin Franklin brought back a purse from Canada in 1724, made from asbestos spun and woven by the Indians. The deposits of asbestos in Canada were discovered in 1860; since then Canada has become a most important source of the fibre.

Today, asbestos is the raw material of an impressive industry, with the fibre serving in many invaluable ways.

Asbestos is the name given to several natural minerals which occur in a fibrous crystalline form. There are three important minerals of this type:

    (a) Anthophyllite
    (b) Amphibole
    (c) Serpentine

A. *Anthophyllite*

This is a magnesium-iron silicate which occurs in some districts in the form of thin plates or fibres. It is not of great importance as a source of commercial asbestos.

B. *Amphibole*

There are several important varieties of this mineral:

(1) *Tremolite* is a calcium magnesium silicate which occurs as greyish masses of brittle, fibrous crystals.

(2) *Actinolite* is an iron calcium magnesium silicate which occurs as greenish masses of fibrous crystals.

(3) *Crocidolite* (Blue Asbestos; Amosite) is a iron sodium silicate which occurs as long flexible fibres of bluish colour, and with a characteristic silky lustre. The fibres are 7.5–10cm (3–4 in). Crocidolite has a higher tensile strength than other asbestos fibres, but is not so resistant to high temperatures as crysotile.

(4) *Mountain Leather* (Mountain Cork) is found in the form of leathery sheets or masses of matted fibres.

(5) *Amphibole Asbestos* (Horneblende Asbestos) is found as fine fibrous crystals, usually of greenish colour.

C. *Serpentine*

This is a hydrated silicate of magnesium which occurs in two fibrous forms:

(1) *Chrysotile* is found as narrow veins in serpentine rock. It is of green to brown colour, and is separated easily into fine silky fibres. This material provides the major part of the world's supply of asbestos.

(2) *Picrolite* is found as fibrous masses in serpentine rock. The fibrous crystals may be 33–36cm (13–14 in). They are inflexible and difficult to separate without breaking.

The bulk of the asbestos used today comes from Canada, South Africa, Rhodesia and Russia. Canada produces more than two-thirds of the total supply.

Opening

The compressed mass of fibrous crystals forming raw asbestos is given a preliminary crushing as the first stage of the opening process. A rotating wheel and pan crusher may be used, great care being taken to avoid undue breakage of the brittle fibres.

From the crusher, the asbestos passes through further opening machines, generally of the toothed roller type. Dirt and powdered rock are removed by means of grids. When the fibrous mass has been opened thoroughly, it may be blended with cotton or other fibres before passing to the carding machine.

Carding

Asbestos fibre is carded by combing with rotating brushes covered with steel bristles. Short fibres and impurities are removed, and the longer asbestos fibres are delivered as a loose web or sheet of fibre. As it leaves the carding machine, the sheet is split into narrow ribbons or rovings, which are wound onto spools.

Spinning

Asbestos rovings are spun on conventional spinning frames (ring or flyer). Doubling is commonly used to increase the uniformity of the yarn.

Asbestos may be mixed with other fibres before spinning, and yarns may also be spun round cores of cotton, glass, nylon, metal wire and other materials.

Short asbestos fibres may be spun by first beating them with water, and then feeding the pulp into a paper machine. A thin sheet of asbestos paper is formed, which is dried and cut into strips which are twisted into yarn.

STRUCTURE AND PROPERTIES

Fine Structure and Appearance

The fibrous crystals of commercial asbestos are 12–300mm (½–12 in) or more in length. They have a smooth, regular surface resembling that of a glass fibre. They are usually near-circular or polygonal in cross-section.

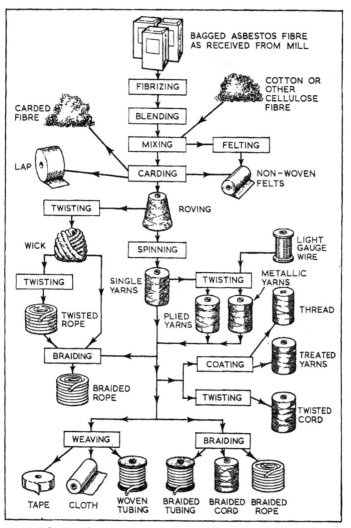

BAGGED ASBESTOS FIBRE
AS RECEIVED FROM MILL

FIBRIZING

BLENDING

COTTON OR
OTHER
CELLULOSE
FIBRE

CARDED
FIBRE

MIXING → FELTING

LAP

CARDING → NON-WOVEN
FELTS

TWISTING ← ROVING

WICK

SPINNING

LIGHT
GAUGE
WIRE

TWISTING

SINGLE
YARNS

TWISTING

METALLIC
YARNS

TWISTED
ROPE

PLIED
YARNS

THREAD

BRAIDING

COATING

TREATED
YARNS

BRAIDED
ROPE

TWISTING

TWISTED
CORD

WEAVING

BRAIDING

TAPE   CLOTH   WOVEN
TUBING

BRAIDED
TUBING

BRAIDED
CORD

BRAIDED
ROPE

Asbestos Fibre Flow Chart – After Asbestos Textile Institute

There is virtually no limit to the fineness of asbestos fibres; the crystals may be subdivided until they are so fine that they cannot be seen through the optical microscope. The molecules of asbestos minerals are arranged in the form of curved sheets which build up into cylindrical structures or tubes. The fibrous crystal of asbestos is made up of many of these tubular structures held together by a mass of crystalline material. The tubes will separate easily one from another, enabling the fibrous crystals to be split into finer and finer fibres.

Other Properties

The outstanding property of all commercial asbestos fibres is their resistance to heat and burning. They are also highly resistant to acids, alkalies and most common chemicals.

Asbestos does not deteriorate in normal use, and it is not attacked by micro-organisms or insects.

ASBESTOS IN USE

Asbestos yarn is durable but has little strength. It is woven into tapes, cloth, brake linings, gaskets and twine. The main applications are those in which its resistance to heat and burning are all-important, such as conveyor belting for hot materials, industrial packings and gaskets, fireproof clothing, theatre curtains and scenery, electrical windings and installation.

DIRECTORY OF
NATURAL FIBRES

ABACA
Leaf; *Musa textilis* (*see* page 30).

ABUTILON AVCENNAE
See *Chinese Jute*.

ABUTILON LONGICUSPE
See *Zada buack*.

ABUTILON PERIPLOCIFOLIUM
See *Maholtine*.

ABUTILON THEOPHRASTI
See *Chinese Jute, Chingma, Tientsin Jute*.

ACROCOMIA SPECIES
See *Corojo, Pita de Corojo*.

AECHME MAGDALENAE
See *Arghan, Bromelia magdalenae, Bromelia longissima, Pineapple (Wild), Pita Floja (Pita Floya), Silk Grass*.

AFRICAN FIBRE
Palm leaf; *Chamaerops humilis*.

AFRICAN SISAL
Leaf; *Agave sisalana*.

AGAVE CANTALA
See *Cantala, Cebu Maguey, Java Cantala, Maguey, Manila Maguey, Nanas Sabrong, Philippine Maguey, Poepoes*.

AGAVE COCUI
See *Dispopo*.

AGAVE DEWEYANA
See *Zapupe larga*.

AGAVE FALCATA
See *Guapilla*.

AGAVE FOURCROYDES
See *Cuban Sisal, Henequen, Mexican Sisal, Sisal Weisz, Victoria Sisal, Yucatan Sisal*.

AGAVE FUNKIANA
See *Istle Jaumave, Jaumave, Jaumave Istle, Jaumave Lechiguilla, Tampico*.

AGAVE HEXAPETALA
See *Cocuiza*.

AGAVE LESPINASSEI
See *Zapupe Fuerte*.

AGAVE LETONAE
See *Letona, Salvador Henequen, San Salvador Sisal*.

AGAVE LOPHANTHA
See *Istle Tula, Lechuguilla, Tampico, Tula Istle*.

AGAVE PALMERI
See *Chino Bermejo, Mano Largo*.

AGAVE PES-MULAE
See *Pata de Mula, Pie de Mula*.

AGAVE PSEUDOTEQUILANA
See *Mescal*.

AGAVE SISALANA
See *African Sisal, Haitian Sisal, Java Sisal, Sisal, Soudan Sisal, Yacci* (*Yaxi*).

AGAVE STRIATA
See *Espadinin*.

AGAVE TEQUILANA
See *Chino Azul, Mazatlan Hemp, Mescal, Mescal Maguey, Tequila*.

AGAVE ZAPUPE
See *Zapupe*.

AKE-IRE
Bast; *Urena lobata*.

AKUND
Seed; *Calotropis gigantea* and *C. procera*. Also *Calotropis Floss* (*see* page 75).

ALFA
Grass leaf; *Stipa tenacissima*.

ALOES CREOLE
Leaf; *Furcraea gigantea*.

ALOES FIBRE
Leaf; *Furcraea gigantea*.

ALOES MALGACHE
Leaf; *Furcraea gigantea.*

ALPACA WOOL
Animal hair; alpaca or *Lama glama* (*see* page 139).

AMBARI
Bast; *Hibiscus cannabinus.*

AMERICAN HEMP
See *Hemp.*

AMERICAN RING TAIL
Animal hair; *Bassariscus astutus.*

ANANAS COMOSUS
See *Pina, Pineapple Fibre.*

ANGORA RABBIT HAIR
Animal hair; Angora rabbit (*see* page 142).

APOCYNUM CANNABINUM
See *Indian Hemp* (United States), *Kendyr.*

APOCYNUM SP.
See *Kendyr.*

ARAMINA
Bast; *Urena lobata.*

AREN
Palm fibre; *Arenga pinnata.*

ARENGA PINNATA
See *Aren, Caba Negro, Ejoo, Gemuti, Palm Fibre* (Capo Negro).

ARGENTINE FLAX
See *Flax.*

ARGHAN
Leaf; *Aechme magdalenae.*

ASBESTOS
Mineral fibre (*see* page 166).

ASCLEPIAS
Milkweed. Bast fibres produced by various species.

ASCLEPIAS INCARNATA
See *Swamp Milkweed.*

175

ASTROCARYUM TUCUMA
See *Tecum, Tucum.*

ATTALEA FUNIFERA
See *Bahia Bass, Bahia Piassava, Bass, Piassava.*

AUSTRIAN FLAX
See *Flax.*

AUSTRIAN HEMP
See *Hemp.*

AWASTHE (AWASTHE HEMP)
Bast; *Hibiscus cannabinus.*

BACTRIS SETOSA
See *Tecum, Tucum.*

BACTRIS SPECIES
See *Mocoro, Tecum, Tucum.*

BADGER
Animal hair; *Meles meles.*

BAHIA BASS
Palm leaf; *Attalea funifera.*

BAHIA PIASSAVA
Palm leaf; *Attalea funifera.*

BALSA FIBRE
Seed; *Ochroma pyramidale* (*see* page 74).

BALTIC FLAX
See *Flax.*

BAMBOO
Stem segments; *Bambusa* species.

BAMBUSA SPECIES
See *Bamboo.*

BAMIA
Bast; *Urena lobata.*

BANANA
Leaf; *Musa sapientum.*

BANANA YUCCA
Leaf; *Yucca mohavensis.*

BAN OCHRA
Bast; *Urena lobata.*

BARIALA
Bast; *Sida micrantha.*

BARRETA
Leaf; *Samuela carnerosana.*

BASS
Palm fibre; *Attalea funifera, Leopoldinia piassaba, Raphia* species, *Vonitra* species.

BASSINE
Palm leaf-stem; *Borassus flabellifer.*

BEAR GRASS
Leaf; *Nolina* species and *Yucca* species, e.g. *Y. angustifolia* and *Y. glauca.*

BEAUMONTIA GRANDIFLORA
Seed (*see* page 74).

BELGIAN FLAX
See *Flax.*

BENARES HEMP
Bast; *Crotalaria juncea.*

BENARES SUNN
Bast; *Crotalaria juncea.*

BILLBERGIA INFUSCATA
See *Infuscata.*

BIMLI PATAM (BIMLI JUTE)
Bast; *Hibiscus cannabinus.*

BLACK SABLE
See *Skunk, Spotted.*

BOEHMERIA NIVEA
See *China Grass, Ramie, Rhea.*

BOHEMIAN-MORAVIAN FLAX
See *Flax.*

BOLO-BOLO
Bast; *Urena lobata.*

BOMBAX COTTON
See *Tree Cotton* (*see* page 73).

BOMBAY ALOE
Leaf; *Furcraea gigantea*.

BOMBAY HEMP
Bast; *Crotalaria juncea* (*see* page 18).

BORASSUS FLABELLIFER
See *Bassine*, *Palmyra Bassine*.

BOWSTRING HEMP
Leaf; *Sansevieria* species.

BRAZILIAN FLAX
See *Flax*.

BRAZILIAN JUTE
Bast; *Corchorus* species.

BROMELIA
Leaf; various species of *Bromelia* (*see* page 34).

BROMELIA KARATAS
See *Gravata*.

BROMELIA LACINIOSA
See *Macambira*.

BROMELIA LONGISSIMA
Leaf; *Aechme magdalenae*.

BROMELIA MAGDALENAE
Leaf; *Aechme magdalenae*.

BROMELIA SAGENARIA
Leaf. Also *Pseudoananas sagenarius*.

BROOM CORN
Flower head; *Sorghum bicolor* (*S. technicum*).

BROOM FIBRE
Bast; *Cytisus scoparius* or *Spartium junceum*.

BROOM ROOT
Root; *Muhlenbergia macoura*.

BROWN HEMP
Bast; *Crotalaria juncea*.

BULGARIAN FLAX
See *Flax*.

BULGARIAN HEMP
See *Hemp*.

BUNTAL
Palm fibre; *Corypha utan*.

BURI
Palm fibre; *Corypha utan*.

CABA NEGRO
Palm fibre; *Arenga pinnata*.

CABULLA
Leaf; *Furcraea cabuya cabuya* and *F. cubensis*.

CABUYA (CABUYA BLANCA, CABUYA BLANCHO,
CABUYA SIN ESPINA)
Leaf; *Furcraea cabuya* and *F. macrophylla* (*see* page 34).

CADILLO
Bast; *Urena lobata*.

CAESAR WEED
Bast; *Urena lobata*.

CALAMUS SPECIES
See *Rattan*.

CALOTROPIS FLOSS
See *Akund* (*see* page 75).

CAMEL HAIR
Animal hair; *Camelus dromedarius* or *Camelus bactrianus* (*see* page 135).

'CAMEL HAIR'
See *Squirrel*.

CANADIAN FLAX
See *Flax*.

CANAMO
Spanish word for hemp.

CANDILLA (CANDILLO)
Bast; *Urena lobata*.

CANHAMO
Bast; *Urena lobata.*

CANNABIS SATIVA
See *American Hemp, Austrian Hemp, Bulgarian Hemp, Chilean Hemp, Chinese Hemp, French Hemp, German Hemp, Hemp, Hungarian Hemp, Illinois Hemp, Italian Hemp, Kentucky Hemp, Manchurian Hemp, Polish Hemp, Roumanian Hemp, Russian Hemp, Russian Siretz, Syrian Hemp, Wisconsin Hemp, Yugoslavian Hemp.*

CANTALA
Leaf; *Agave cantala* (*see* page 33).

CANTON
Leaf; plant of *Musa* species similar to *M. textilis* (*see* page 33).

CARAUA
See *Caroa.*

CARLUDOVICA PALMATA
See *Palm Hat Plant, Panama Hat Palm.*

CAROA
Leaf; *Neoglaziova variegata* (*see* page 34).

CARRAPICHO
Bast; *Urena lobata.*

CARYOTA URENS
See *Kittool.*

CASHMERE
Animal hair; cashmere goat, *Capra* sp. (*see* page 136).

CATTAIL FIBRE
Seed; *Typha latifolia* and *T. angustifolia* (Cattails) (*see* page 75).

CATTLE HAIR
Animal hair; *Bos taurus.*

CEBU HEMP
Leaf; *Musa textilis.*

CEBU MAGUEY
Leaf; *Agave cantala.*

CEIBA AESCULIFOLIA
See *Pochote.*

CEPHALONEMA SPECIES
See *Punga.*

CHAMAEROPS HUMILIS
See *African Fibre, Crin Vegetal.*

CHANVRE
French word for hemp.

CHAPARRAL YUCCA
Leaf; *Hesperoyucca whipplei.*

CHILEAN FLAX
See *Flax.*

CHILEAN HEMP
See *Hemp.*

CHINA GRASS
Bast; *Boehmeria nivea (see* page 22).

CHINESE FAN PALM
Palm leaf segments; *Livistona chinensis.*

CHINESE HEMP
See *Hemp.*

CHINESE JUTE
Bast; *Abutilon theophrasti* and *A. avcennae.*

CHINESE MAT RUSH
Stem; *Lepironia mucronata.*

CHINGMA
Bast; *Abutilon theophrasti.*

CHINO AZUL
Leaf; *Agave tequilana.*

CHINO BERMEJO
Leaf; *Agave palmeri.*

CHIQUE-CHIQUE
Bast; *Leopoldinia piassaba.*

CHORISIA SPECIOSA
Seed *(see* page 74).

CHUCHAO
Leaf; *Furcraea andina.*

CIVET
See *Skunk, Spotted.*

COCONADA HEMP
Bast; *Crotalaria juncea.*

COCONUT FIBRE
Nut husk; *Cocos nucifera.*

COCOS NUCIFERA
See *Coconut Fibre, Coir.*

COCUIZA
Leaf; *Furcraea geminispina.*

COCUIZA
Leaf; *Agave hexapetala; Furcraea geminispina* and *F. humboldtiana.*

COCUIZA MANSA
Leaf; *Furcraea gigantea.*

COIR
Nut husk; *Cocos nucifera (see* page 73).

CONGO JUTE
Bast; *Urena lobata.*

CORCHORUS CAPSULARIS
See *Jute, White Jute.*

CORCHORUS OLITORIUS
See *Daisee Jute, Tossa.*

CORCHORUS SPECIES
See *Brazilian Jute, Daisee Jute, Jute, Tossa, White Jute.*

COROJO
Palm leaf; *Acrocomia* species.

CORYPHA UTAN
See *Buntal, Buri, Palm Fibre (Buri), Raffia.*

COTTON
Seed; species of *Gossypium (see* page 35).

COURTRAI FLAX
See *Flax.*

COUSIN ROUGE
Bast; *Urena lobata.*

CRIN VEGETAL
Palm leaf; *Chamaerops humilis.*

CROTALARIA JUNCEA
See *Benares Hemp, Benares Sunn, Bombay Hemp, Brown Hemp, Coconada Hemp, Indian Hemp, Itarsi Hemp, Jubblepore Hemp, Madras Hemp, Phillibit Hemp, Seonie Hemp, Sunn, Warangel Hemp.*

CUBAN JUTE
Bast; various species of *Urena, Malva, Sida.*

CUBAN SISAL
Leaf; *Agave fourcroydes (see* page 29).

CULUT CULUTAN
Bast; *Urena lobata.*

CYTISUS SCOPARIUS
See *Broom Fibre, Spanish Broom.*

DA
Bast; *Hibiscus cannabinus.*

DAISEE JUTE
Bast; *Corchorus olitorius.*

DAVAO HEMP
Leaf; *Musa textilis.*

DECCAN HEMP
Bast; *Hibiscus cannabinus.*

DHA
Bast; *Hibiscus cannabinus.*

DISPOPO
Leaf; *Agave cocui.*

DUM
Palm leaf; *Hyphaene thebaica.*

DUTCH FLAX
See *Flax.* Also Blue Dutch Flax, White Dutch Flax.

EEL GRASS
Leaf; *Zostera marina.*

EGYPTIAN FLAX
See *Flax.*

EIRE FLAX
See *Flax*.

EJOO
Palm fibre; *Arenga pinnata*.

ENSETE EDULIS
Leaf; *Musa ensete*. Also *Scioa*.

ESCOBILLA
Bast; various species of *Sida*.

ESPADININ
Leaf; *Agave striata*.

ESPARTO
Grass leaf; *Stipa tenacissima*.

ESTHONIAN FLAX
See *Flax*.

FIQUE
Leaf; *Furcraea macrophylla* (*see* page 34).

FITCH
Animal hair; skunk, *Mephitis mephitis* et al.

FLAX
Bast; *Linum usitatissimum* (*see* page 4).

FLEMISH FLAX
See *Flax*.

FORMIO
Leaf; *Phormium tenax*.

FOX
Animal hair; *Vulpes fulva*.

FRENCH FLAX
See *Flax*.

FRENCH HEMP
See *Hemp*.

FUR FIBRES
Animal hair; various species of rabbit (*see page* 141).

FURCRAEA ANDINA
See *Chuchao*.

FURCRAEA CABUYA
See *Cabulla, Cabuya, Cabuya Blanca, Cabuya Blancho, Cabuya Sin Espina.*

FURCRAEA CUBENSIS
See *Cabulla, F. hexapetala.*

FURCRAEA GEMINISPINA
See *Cocuiza.*

FURCRAEA GIGANTEA
See *Aloes Creole, Aloes Fibre, Aloes Malgache, Bombay Aloe, Cocuiza Mansa, Mauritius Hemp, Natal Hemp, Piteira.*

FURCRAEA HEXAPETALA
See *F. cubensis, Maguey* (Cüba), *Peetray, Pitre.*

FURCRAEA HUMBOLDTIANA
See *Cocuiza.*

FURCRAEA MACROPHYLLA
See *Cabuya, Fique, Maguey* (Peru).

GALGAL
See *Kumbi* (*see* page 74).

GALLA
Leaf; *Musa ensete.*

GAMBO HEMP (GOMBO HEMP)
Bast; *Hibiscus cannabinus.*

GEMUTI
Palm fibre; *Arenga pinnata.*

GENET
Animal hair; *Genetta.*

GERMAN FLAX
See *Flax.*

GERMAN HEMP
See *Hemp.*

GOAT HAIR
Animal hair; various species of *Capra.*

GRAND COUSIN
Bast; *Urena lobata.*

GRAND MAHOT COUSIN
Bast; *Urena lobata.*

GRAVATA
Leaf; *Bromelia karatas.*

GUANACO
Animal hair; guanaco.

GUAPILLA
Leaf; *Agave falcata.*

GUAXIMA
Bast; *Urena lobata* (includes *Guaxima roxa* and *G. vermehla*).

GUIAZO
Bast; *Urena lobata.*

GUINEA HEMP
Bast; *Hibiscus cannabinus* (*see* page 20).

HAITIAN SISAL
Leaf; *Agave sisalana.*

HANF
German word for hemp.

HEMP
Bast; *Cannabis sativa* (*see* page 17).

HENEQUEN
Leaf; *Agave fourcroydes* (*see* page 29).

HENNUP
Dutch word for hemp.

HESPERALOE FUNIFERA
See *Ixtli, Samandoca, Zamandoque.*

HESPEROYUCCA WHIPPLEI
Leaf; see *Chaparral Yucca, Yucca.*

HIBISCUS ABELMOSCHUS
See *Musk Hemp.*

HIBISCUS CANNABINUS
See *Ambari, Awasthe (Awasthe Hemp), Bimli Patam (Bimli Jute), Deccan Hemp, Da, Dha, Gambo Hemp (Gombo Hemp), Guinea Hemp, Kenaf, Meshta.*

HIBISCUS ESCULENTUS
See *Ochra*.

HIBISCUS FERAX
See *Meshta, Papoula de St Francis, St Francis Poppy*.

HIBISCUS KITAIBELIFOLIUS
See *Juta Paulista*.

HIBISCUS RADIATUS
See *Papoula de St Francis*.

HIBISCUS SABDARIFFA
See *Java Jute, Roselle (Rosella)*.

HOG BRISTLE
Animal hair; *Sus scrofa*.

HORSE HAIR
Animal hair; *Equus caballus*.

HUARIZO
Animal hair; cross between llama sire and alpaca dam (*see* page 140).

HUNGARIAN FLAX
See *Flax*.

HUNGARIAN HEMP
See *Hemp*.

HYPHAENE THEBAICA
See *Dum*.

IFE HEMP (IFÉ)
Leaf; *Sansevieria* species.

ILLINOIS HEMP
See *Hemp*.

INDIAN HEMP
Bast fibres from several unrelated plants, including species of *Crotalaria* and *Hibiscus*, e.g. *Ambari, Benares Hemp, Itarsi, Sunn* (*see* page 18).

INDIAN HEMP (UNITED STATES)
Bast; *Apocynum cannabinum*.

INFUSCATA
Leaf; *Billbergia infuscata*.

187

IRISH FLAX
See *Flax*.

ISTLE; IXTLE
Generic Mexican term for various species of *Agave*.

ISTLE JAUMAVE
Leaf; *Agave funkiana*.

ISTLE, PALMA
Leaf; *Samuela carnerosana*.

ISTLE PITA
Leaf; *Yucca treculeana*.

ISTLE TULA
Leaf; *Agave lophantha*.

ITALIAN FLAX
See *Flax*.

ITALIAN HEMP
See *Hemp*.

ITARSI HEMP
Bast; *Crotalaria juncea*.

IXTLI
Leaf; *Hesperaloe funifera*.

JAUMAVE; JAUMAVE ISTLE; JAUMAVE LECHIGUILLA
Leaf; *Agave funkiana*.

JAVA CANTALA
Leaf; *Agave cantala*.

JAVA JUTE
Bast; *Hibiscus sabdariffa*.

JAVA KAPOK
Seed; *Ceiba pentandra*. See *Kapok* (page 74).

JAVA SISAL
Leaf; *Agave sisalana*.

JIRICA
Leaf; *Nolina* species.

JUBBLEPORE HEMP
Bast; *Crotalaria juncea*.

JUTA PAULISTA
Bast; *Hibiscus kitaibelifolius.*

JUTE
Bast; *Corchorus capsularis* and *C. olitorius* (*see* page 13).

JUTILAL
Typha species.

KAPOK
Seed; *Ceiba pentandra.*

KENAF
Bast; *Hibiscus cannabinus* (*see* page 20).

KENDYR
Bast; *Apocynum sp.*

KENTUCKY HEMP
See *Hemp.*

KITTOOL
Palm fibre; *Caryota urens.*

KOLINSKY
See *Sable, Red.*

KUMBI
Seed; *Cochlospermum gossypium.* Also *Galgal* (*see* page 74).

KUNJIA
Bast; *Urena sinuata.*

LATVIAN FLAX
See *Flax.*

LECHUGUILLA
Leaf; *Agave lophantha.*

LEOPOLDINIA PIASSABA
See *Bass, Chique-Chique, Monkey Bass, Para Piassava.*

LEPIRONIA MUCRONATA
See *Chinese Mat Rush.*

LETONA
Leaf; *Agave letonae* (*see* page 33).

LINEN
See *Flax.*

LINUM USITATISSIMUM
See *Argentine Flax, Austrian Flax, Baltic Flax, Belgian Flax, Bohemian–Moravian Flax, Brazilian Flax, Bulgarian Flax, Canadian Flax, Chilean Flax, Courtrai Flax, Dutch Flax, Egyptian Flax, Eire Flax, Esthonian Flax, Flax, Flemish Flax, French Flax, German Flax, Hungarian Flax, Irish Flax, Italian Flax, Latvian Flax, Linen, Lithuanian Flax, Livonian Flax, Manchurian Flax, New Zealand Flax, Northern Ireland Flax, Oregon Flax, Peruvian Flax, Polish Flax, Roumanian Flax, Russian Flax, Yugoslavian Flax.*

LITHUANIAN FLAX
See *Flax.*

LIVISTONA CHINENSIS
See *Chinese Fan Palm.*

LIVONIAN FLAX
See *Flax.*

LLAMA WOOL
Animal hair; *Lama glama* (*see* page 138).

LOOFAH
Leaf; *Luffa* species.

LUFFA
See *Loofah, Luffa Gourd, Vegetable Sponge.*

LUFFA GOURD
Leaf; *Luffa* species.

MACAMBIRA
Leaf; *Bromelia laciniosa.*

MADAGASCAR BASS
Vonitra species.

MADRAS HEMP
Bast; *Crotalaria juncea.*

MAGUEY
Leaf; *Agave cantala* (*see* page 33).

MAGUEY
Latin American term used for various species of *Agave.*

MAGUEY (CUBA)
Leaf; *Furcraea hexapetala.*

MAGUEY (PERU)
Leaf; *Furcraea macrophylla.*

MAHOLTINE
Bast; *Abutilon periplocifolium.*

MALACHE MALACOPHYLLA
See *Malva velluda.*

MALVA
Latin American term applied to various species of *Malvaceae.*

MALVA
Bast; *Sida micrantha.*

MALVA BLANCA
Bast; *Urena lobata.*

MALVA LISTRO
See *Malva.*

MALVA RELLUDO
Bast; *Pavonia* species and *Uacima.*

MALVA RISCO
See *Malva.*

MALVA ROXA
Bast; *Urena lobata.*

MALVAS
Bast; *Urena lobata.*

MALVA VELLUDA
Bast; *Malache malacophylla.*

MANCHURIAN FLAX
See *Flax.*

MANCHURIAN HEMP
See *Hemp.*

MANILA
Leaf; *Musa textilis.*

MANILA HEMP
Leaf; *Musa textilis* (*see* page 30).

MANILA MAGUEY
Leaf; *Agave cantala.*

MANO LARGO
Leaf; *Agave palmeri*.

MAURITIA FLEXUOSA
See *Moriche Palm, Palm Fibre (Nirucge)*.

MAURITIUS HEMP (MAURITIUS FIBRE)
Leaf; *Furcraea gigantea (see page 33)*.

MAZATLAN HEMP
Leaf; *Agave tequilana*.

MESCAL
Leaf; *Agave tequilana* and various species of *Agave*.

MESCAL MAGUEY
Leaf; *Agave tequilana*.

MESHTA
Bast; *Hibiscus ferax*.

MESHTA
Bast; *Hibiscus cannabinus*.

MEXICAN RICE ROOT
See *Zacaton*.

MEXICAN SISAL
Leaf; *Agave fourcroydes*.

MILKWEED FIBRE
Bast; *Asclepias* species. See *Swamp Milkweed*. (Note: milkweed also produces seed hairs used as fibres – see *Milkweed Floss*.)

MILKWEED FLOSS
Seed; various species of *Asclepias (see page 75)*.

MINK
See *Sable, Red*.

MOCORO
Palm leaf; *Bactris* species.

MOHAIR
Animal hair; Angora goat, *Capra (see page 132)*.

MONKEY BASS
Palm leaf-stem; *Leopoldinia piassaba*.

MOORVA
Leaf; *Sansevieria* species.

MORICHE PALM
Palm leaf; *Mauritia flexuosa.*

MUHLENBERGIA MACOURA
See *Broom Root, Rice Root.*

MUSA ENSETE
See *Ensete Edulis, Galla, Sidamo.*

MUSA SAPIENTUM
See *Banana.*

MUSA TEXTILIS
See *Abaca, Cebu Hemp, Davao Hemp, Manila, Manila Hemp.*

MUSK HEMP
Bast; *Hibiscus abelmoschus.*

MUSKRAT, NORTHERN
Animal hair; *Ondatra zibethicus.*

MUSKRAT, SOUTHERN
Animal hair; *Ondatra rivalicia.*

NANAS SABRONG
Leaf; *Agave cantala.*

NATAL HEMP
Leaf; *Furcraea gigantea.*

NEOGLAZIOVIA VARIEGATA
See *Caraua, Caroa.*

NETTLE
Bast; *Urtica dioica, U. urens, U. pilulifera (see* page 25)*.*

NEW ZEALAND FLAX
See *Flax.* (This is true flax, not to be confused with *Phormium,* also commonly called New Zealand Flax.)

NEW ZEALAND FLAX (NEW ZEALAND HEMP)
Leaf; *Phormium tenax (see* page 33)*.*

NOLINA SPECIES
See *Bear Grass, Jirica, Scaahuista.*

NORTHERN IRELAND FLAX
See *Flax.*

OCHRA
Bast; *Hibiscus esculentus.*

OLONA
Bast; *Touchardia latifolia.*

OREGON FLAX
See *Flax.*

OTOTE GRANDE
Bast; *Urena lobata.*

OX HAIR
Animal hair; *Bos taurus.*

OZONE FIBRE
Seed; *Asclepias sp. incarnata.*

PACOL
Leaf; plant of *Musa* species (*see* page 33).

PAKA
Bast; *Urena lobata.*

PALMA
Leaf; *Samuela carnerosana* (*see* page 34).

PALMA BARRETA
Leaf; *Samuela carnerosana.*

PALMA ISTLE
Leaf; *Samuela carnerosana.*

PALMETTO
Palm fibre; *Sabal* species.

PALM FIBRE (BURI)
Palm fibre; *Corypha utan.*

PALM FIBRE (CAPO NEGRO)
Palm fibre; *Arenga pinnata.*

PALM FIBRE (NIRUCGE)
Palm leaf; *Mauritia flexuosa.*

PALM HAT PLANT
Leaf segments; *Carludovica palmata.*

PALMILLA
Leaf; *Yucca elata.*

PALMYRA
Palm leaf-stem; *Borassus flabellifer.*

PALMYRA BASSINE
Palm leaf-stem; *Borassus flabellifer.*

PANAMA HAT PALM
Leaf segments; *Carludovica palmata.*

PANGANE (PANGANE HEMP)
Leaf; *Sansevieria Kirkii.*

PAPOULA DE ST FRANCIS
Bast; *Hibiscus ferax* and *H. radiatus.*

PARA PIASSAVA
Palm leaf-stem; *Leopoldinia piassaba.*

PATA DE MULA
Leaf; *Agave pes-mulae.*

PAVONIA SPECIES
P. malacophylla, *P. schimperiana*, *P. tomentosa.* See *Malva relludo, Uacima.*

PEETRAY
See *Pitre.*

PERSIAN GOAT HAIR
Animal hair; Persian goat (*see* page 138).

PERUVIAN FLAX
See *Flax.*

PHILLIBIT BLACK HEMP
Bast; *Crotalaria juncea.*

PHILIPPINE MAGUEY
Leaf; *Agave cantala.*

PHORMIUM
Leaf; *Phormium tenax* (*see* page 33).

PHORMIUM TENAX
See *Formio, New Zealand Flax, New Zealand Hemp, Phormium, St Helena Hemp.*

PIASSAVA
Palm fibre; *Attalea funifera, Leopoldinia piassaba, Raphia gigantea, Vonitra* species (*see* page 34).

PIE DE MULA
Leaf; *Agave pes-mulae.*

PINA
Leaf; *Ananas comosus.*

PINEAPPLE FIBRE
Leaf; *Ananas comosus (see* page 34).

PINEAPPLE WILD
Leaf; *Aechme magdalenae.*

PITA
Latin American name for various hard fibres.

PITA DE COROJO
Palm leaf; *Acrocomia* species.

PITA FLOJA (PITA FLOYA)
Leaf; *Aechme magdalenae (see* page 34).

PITA ISTLE
Leaf; *Yucca treculeana.*

PITA PALMA
Leaf; *Yucca treculeana.*

PITEIRA
Leaf; *Furcraea gigantea.*

PITRE
Leaf; *Furcraea hexapetala.*

POCHOTE
Seed; *Ceiba aesculifolia.*

POEPOES
Leaf; *Agave cantala.*

POLISH FLAX
See *Flax.*

POLISH HEMP
See *Hemp.*

POLOMPOM
Bast; *Thespesia lampas* and *T. populnea.*

PONY
Animal hair; *Equus caballus.*

PSEUDOANANAS SAGENARIUS
Leaf. Also *Bromelia sagenaria* and *Soft Leaf Fibre*.

PUNGA
Bast; *Cephalonema* species.

QUEENSLAND HEMP
Bast; *Sida rhombifolia*.

RABBIT, COMMON
Animal hair; *Oryctolagus cuniculus* (*see* page 141).

RAFFIA
Palm leaf segments; *Corypha utan, Raphia ruffia, Raphia vinifera* (*see* page 35).

RAMIE
Bast; *Boehmeria nivea* (*see* page 22).

RAPHIA GIGANTEA
See *Piassava*.

RAPHIA RUFFIA
See *Bass, Raffia*.

RAPHIA VINIFERA
See *Bass, Raffia, West African Bass*.

RATTAN
Stem; *Calamas* species.

RHEA
Bast; *Boehmeria nivea* (*see* page 22).

RICE PAPER PLANT
Pith; *Tetrapanax papyriferus*.

RICE ROOT
Root; *Muhlenbergia macoura*.

ROSELLE (ROSELLA)
Bast; *Hibiscus sabdariffa*.

ROUMANIAN FLAX
See *Flax*.

ROUMANIAN HEMP
See *Hemp*.

RUSSIAN FLAX
See *Flax*.

RUSSIAN HEMP
See *Hemp*.

RUSSIAN SIRETZ
See *Hemp*.

SABAL SPECIES
See *Palmetto*.

SABLE, RED
Animal hair; *Mustela sibirica*. Also *Kolinsky, China Mink, Jap Mink*.

SALVADOR HENEQUEN
Leaf; *Agave letonae*.

SAMANDOCA
Leaf; *Hesperaloe funifera*.

SAMCHU
Seed; *Chorisia* species.

SAMUELA CARNEROSANA
See *Barreta, Istle (Palma), Palma, Palma Barreta, Palma Istle*.

SANN
See *Sunn*.

SAN SALVADOR SISAL
Leaf; *Agave letonae*.

SANSEVIERIA
Leaf; various species of *Sansevieria*, e.g. *S. Kirkii* (Pangane) and
S. trifasciata (*see* page 34).

SANSEVIERIA CYLINDRICA
See *Ifé*.

SANSEVIERIA KIRKII
See *Pangane*.

SANSEVIERIA SPECIES
See *Bowstring Hemp, Ifé Hemp, Ifé, Moorva, Pangane, Pangane
Hemp, Sansevieria*.

SCAAHUISTA
Leaf; *Nolina* species.

SCIOA
Leaf; *Ensete edulis.*

SEONIE HEMP
Bast; *Crotalaria juncea.*

SHEEP WOOL
See *Wool.*

SIDA
See *Bariala, Escobilla.*

SIDA MICRANTHA
See *Bariala, Escobilla, Malva, Malva Listro, Malva Risco.*

SIDAMO
Leaf; *Musa ensete.*

SIDA RHOMBIFOLIA
See *Queensland Hemp.*

SILK
Animal protein fibre spun by *Bombyx mori* (*see* page 143).

SILK GRASS
Leaf; *Aechme magdalenae.*

SILK, TUSSAH
Animal protein fibre spun by *Antheraea paphia* et al.

SISAL
Leaf; *Agave sisalana* (*see* page 27).

SISAL WEISZ
Leaf; *Agave fourcroydes.*

SKUNK
See *Fitch.*

SKUNK, SPOTTED
Animal hair; *Spilogale* sp.

SOAP WEED
Leaf; *Yucca angustifolia* and *Y. glauca.*

SOFT LEAF FIBRE
Leaf; *Pseudoananas sagenarius.*

SORGHUM BICOLOR (S. TECHNICUM)
See *Broom Corn.*

SOUDAN SISAL
Leaf; *Agave sisalana.*

SOUTHERN MOSS
Fibrous stem axis; *Tillandsia usneoides.*

SPANISH BROOM
Bast; *Cytisus scoparius.*

SPANISH DAGGER
Leaf; *Yucca macrocarpa.*

SPANISH MOSS
Fibrous stem axis; *Tillandsia usneoides.*

SPARTIUM JUNCEUM
See *Broom Fibre.*

SPIDER SILK
Filaments of animal protein spun by spiders (*see* page 157).

SQUIRREL
Animal hair; *Sciurus vulgaris* et al.

ST FRANCIS POPPY
Bast; *Hibiscus ferax* and *H. radiatus.*

STIPA TENACISSIMA
See *Alfa, Esparto.*

STROPHANTHUS SPP.
Seed (*see* page 74).

SUNN (SANN)
Bast; *Crotalaria juncea* (*see* page 18).

SWAMP MILKWEED
Bast; *Asclepias incarnata.*

SYRIAN HEMP
See *Hemp.*

TAMPICO
Leaf; *Agave funkiana* and *A. lophantha.*

TECUM; TUCUM
Palm leaf; *Astrocaryum tucuma* and *Bactris setosa* (and other *Bactris* species).

TEQUILA
Leaf; *Agave tequilana.*

TETRAPANAX PAPYRIFERUS
See *Rice Paper Plant.*

THESPESIA LAMPAS; T. POPULNEA
See *Polompom.*

TIENTSIN JUTE
Bast; *Abutilon theophrasti.*

TILLANDSIA USNEOIDES
See *Southern Moss, Spanish Moss, Tree Beard.*

TOJA
Bast; *Urena lobata.*

TOSSA
Bast; *Corchorus olitorius.*

TOUCHARDIA LATIFOLIA
See *Olona.*

TREE BEARD
Fibrous stem axis; *Tillandsia usneoides.*

TREE COTTON
Seed; *Malvaceae* species. Also *Vegetable Down, Bombax Cotton* (*see* page 73).

TULA ISTLE
Leaf; *Agave lophantha.*

TUSSAH SILK (TUSSORE SILK; TUSSUR SILK)
Filament spun by various species of silkworm of the genus *Antheraea* (*see* page 151).

TYPHA SPECIES
See *Cat-tail Fibre, Jutilal.*

UACIMA (UAIXIMA)
Bast; *Urena lobata*. Also *Malva Relludo*.

URENA LOBATA
See *Ake-ire, Aramina, Bamia, Ban Ochra, Bolo-Bolo, Cadillo, Caesar Weed, Candilla (Candillo), Canhamo, Carrapicho, Congo Jute, Cousin Rouge, Culut Culutan, Grand Cousin, Grand Mahot Cousin, Guaxima, Guiazo, Malva Blanca, Malva Roxa, Malvas, Otote Grande, Paka, Toja, Uacima (Uaixima)* (*see* page 21).

URENA SINUATA
See *Cuban Jute, Kunjia*.

URTICA NIVEA
See *Boehmeria nivea*.

URTICA SPECIES
See *Nettle (U. dioica, U. urens, U. pilulifera)*.

VEGETABLE DOWN
See *Tree Cotton* (page 73).

VEGETABLE SPONGE
Fibrous net of fruit; *Luffa* species.

VICTORIA SISAL
Leaf; *Agave fourcroydes*.

VICUNA
Animal hair; *Vicugna vicugna* (*see* page 140).

VONITRA SPECIES
See *Bass, Madagascar Bass, Piassava* (Madagascar).

WARANGEL HEMP
Bast; *Crotalaria juncea*.

WEST AFRICAN BASS
Palm leaf-stem; *Raphia vinifera*.

WHITE JUTE
See *Jute*.

WILD SILK
See *Tussore Silk* (page 151).

WISCONSIN HEMP
See *Hemp*.

WOOL
Animal hair; *Ovis aries* et al. (*see* page 80).

YACCI (YAXI)
Leaf; *Agave sisalana*.

YUCATAN SISAL
Leaf; *Agave fourcroydes* (*see* page 29).

YUCCA
Many species yield fibre, e.g. *Y. angustifolia, Y. glauca, Y. treculeana*.
See *Bear Grass, Soap Weed, Pita Istle*. See also *Hesperoyucca whipplei*.

YUCCA ANGUSTIFOLIA
See *Bear Grass, Soap Weed*.

YUCCA ELATA
See *Palmilla*.

YUCCA GLAUCA
See *Bear Grass, Soap Weed*.

YUCCA MACROCARPA
See *Spanish Dagger*.

YUCCA MOHAVENSIS
See *Banana Yucca*.

YUCCA TRECULEANA
See *Istle Pita, Pita Istle, Pita Palma*.

YUGOSLAVIAN FLAX
See *Flax*.

YUGOSLAVIAN HEMP
See *Hemp*.

ZACATON
Root; *Mexican Rice Root*.

ZADA BUACK
Bast; *Abutilon longicuspe*.

ZAMANDOQUE
Leaf; *Hesperaloe funifera.*

ZAPUPE
Leaf; *Agave zapupe.*

ZAPUPE FUERTE
Leaf; *Agave lespinassei.*

ZAPUPE LARGA
Leaf; *Agave deweyana.*

ZOSTERA MARINA
See *Eel Grass.*

INDEX

205

Printed and bound by CPI Group (UK) Ltd, Croydon, CR0 4YY

03/10/2024

01040435-0001